Fiery Temporalities in Theatre and Performance

Methuen Drama Engage offers original reflections about key practitioners, movements and genres in the fields of modern theatre and performance. Each volume in the series seeks to challenge mainstream critical thought through original and interdisciplinary perspectives on the body of work under examination. By questioning existing critical paradigms, it is hoped that each volume will open up fresh approaches and suggest avenues for further exploration.

Series Editors

Mark Taylor-Batty
University of Leeds, UK

Enoch Brater
University of Michigan, USA

Titles

Adaptation in Contemporary Theatre
by Frances Babbage
ISBN 978-1-4725-3142-1

Authenticity in Contemporary Theatre and Performance
by Daniel Schulze
ISBN 978-1-350-00096-4

Beat Drama: Playwrights and Performances of the "Howl" Generation
edited by Deborah R. Geis
ISBN 978-1-4725-6787-1

Drama and Digital Arts Cultures
by David Cameron, Michael Anderson and Rebecca Wotzko
ISBN 978-1-4725-9219-4

Social and Political Theatre in 21st-Century Britain: Staging Crisis
by Vicky Angelaki
ISBN 978-1-4742-1316-5

Theatre in the Dark: Shadow, Gloom and Blackout in Contemporary Theatre
edited by Adam Alston and Martin Welton
ISBN 978-1-4742-5118-1

Watching War On The Twenty-First-Century Stage: Spectacles of Conflict
by Clare Finburgh
ISBN 978-1-4725-9866-0

Fiery Temporalities in Theatre and Performance

The Initiation of History

Maurya Wickstrom

Series Editors
Enoch Brater and Mark Taylor-Batty

methuen | drama
LONDON • NEW YORK • OXFORD • NEW DELHI • SYDNEY

METHUEN DRAMA
Bloomsbury Publishing Plc
50 Bedford Square, London, WC1B 3DP, UK

BLOOMSBURY, METHUEN DRAMA and the Methuen Drama logo
are trademarks of Bloomsbury Publishing Plc

First published in Great Britain 2018
This paperback edition first published 2020

Series design by Louise Dugdale
Cover image: South African artist William Kentridge's
The Refusal of Time at The Metropolitan Museum of Art in New York,
United States, 2014. (© Nardus Engelbrecht/Gallo Images/Getty Images)

A catalogue record for this book is available from the British Library.

Library of Congress Cataloging-in-Publication Data
Names: Wickstrom, Maurya, author.
Title: Fiery temporalities in theatre and performance : the initiation of
history / Maurya Wickstrom.
Description: London : Methuen Drama, 2018. | Series: Methuen drama engage |
Includes bibliographical references and index.
Identifiers: LCCN 2017061313| ISBN 9781474281690 (hb) | ISBN 9781474281713
(epdf) | ISBN 9781474281706 (eBook)
Subjects: LCSH: European drama–History and criticism. | Space and time in
literature. | Space and time in the theater.
Classification: LCC PN1650.S63 W53 2018 | DDC 809.2/9353–dc23 LC record available at
https://lccn.loc.gov/2017061313

ISBN: HB: 978-1-4742-8169-0
PB: 978-1-3501-4329-6
ePDF: 978-1-4742-8171-3
eBook: 978-1-4742-8170-6

Series: Methuen Drama Engage

Typeset by Deanta Global Publishing Services, Chennai, India

To find out more about our authors and books visit
www.bloomsbury.com and sign up for our newsletters.

For Joel Reynolds (1957–2015)
and for our children, Erin and Naoise Reynolds

Contents

Acknowledgments

I wish to thank, in the first place, Alan Read. Not only has he encouraged me in every way in general, he introduced me to Mark Dudgeon on that lovely late night around the pool at the hotel in Palo Alto during the Performance Studies International Conference in 2012. I am grateful to Mark, in turn, for listening with such intelligence and responsiveness to my ideas for this book on that night and on subsequent occasions. He took the project on without much guarantee. I hope the book is worthy of his confidence.

The book was written under difficult circumstances in which there was also much beauty. With regard to this book, that beauty appeared especially in the brilliance and gentleness of two separate sets of graduate students with whom I explored performance and temporality. One was at the CUNY Graduate Center. Thank you so very much Clio, Amir, Bhargav, Seth, Andrew, and the other invaluable members of the follow-up "salon"—Eylül and Ugo. The other was at the Yale School of Drama. Thank you so very much David, Maria, Libby, Josh, Kari, Ariel, and Zizi. I am grateful to Catherine Sheehy for extending to me the invitation to teach Performance and Temporality at Yale, and for her support and kindness then and now. Others at Yale were important to me, especially Alex Ripp, Elise Morrison, and others at the Interdisciplinary Performance Studies Program, where I was able to share work on the Toussaint material at a very early stage. Kimberly Jannarone, thank you.

Thanks very much to Mariellen Sandford and TDR, which has been a home for me as I wrote my first essays on temporal themes, about The TEAM's *Architecting* and then Cassils's *Tiresias*. Portions of Chapter 4, when I extend my thinking about Cassils's work, were previously published there. Thanks to Jill Dolan and Stacy Wolf for extending the initial invitation to me to write for the volume in which the Cassils essay appears, and for their excellent editing. Thanks also to Nicholas Ridout, who invited me to write for his Marxism and Performance issue of *Theatre Survey* and therefore allowed me to explore temporal thought in relation to the work of M. Lamar (and edited it so well). Thanks so very much to Cassils, M. Lamar, Rachel Chavkin, and Andrew Schneider, all of whom were generous in their willingness to be in conversation with me.

At the Graduate Center and beyond, Dr. Rayya El Zein has been a student who has taught me, challenged me, and supported me as I worked on these ideas. Her work on and commitment to Palestine and other places of struggle and the initiation of history is brilliant and commendable. Eylül Akinci, a brilliant, fierce, and compassionate woman and scholar, thank you for everything, and for taking care of my house and my cats so I could finish this. Thank you to the College of Staten Island and George Emilio Sanchez for the leave time to write. My colleague there, Sean Edgecomb, has been an intelligent and indispensable support as I have developed this work. As always, Jennifer Parker-Starbuck has stayed by my side as colleague and friend. Jill Dolan and Stacy Wolf, your ongoing interest and belief has meant and means everything.

Finally, I wish to express my gratitude in general to all of those in the world who are taking time into their hands, initiating history, forming compassionate forms of revolting together, standing at the edge of the to-come.

1

Introduction

This book began in 2012 at the Brooklyn Academy of Music (BAM) as I watched the Israeli-born and -raised choreographer Hofesh Shechter's *Political Mother*. The piece had premiered in London in 2010. In the intervening year, Tahrir Square and Liberty Square were occupied. The almost taste of revolution was elusive and exhilarating. That intervening year was a time of riots and occupations. *Political Mother* 2010 preceded and predicted this time, and *Political Mother* 2012 read back onto it. The performance generated my thoughts on what became the central concerns of this book: the capacity of theatre and performance to invent or inaugurate particular types of time, which allow us to be in history, as history, in a time of our own creation.

This book suggests a politics of time in which temporal ideologies, practices, and habits that work to ensure passivity are subordinated to the energy of fire, seizure, and initiation. Some of the many theatre and performance works in this book are overtly political. All are a practice of politics as, variously, the inauguration of temporalities that are disassociated from the category of the future and from the centrality of death that informs many dominant temporal schemas. The temporalities explored in these pages are those of the new present, penultimate time, and kairos. They are temporal capacities and practices that can allow us to exit chronological time as the internal structure of history as it is usually understood. All of the theatre and performance works in this book, by virtue of their initiation of time, offer the possibility of history felt and lived differently. While not all of the performances in the book directly have to do with history, they all have to do with time. I proceed according to the proposition that any intervention in temporality is the beginning of the possibility of a new form of history.

Political Mother: Introductory Remarks on Chronological Time and Changing Time

I turn here to *Political Mother* to begin the process of entering this work, this book, which will at times become thick with philosophy, but always borne in and through performance. This description will introduce some of the ideas and terms that will figure significantly in the chapters to come. I will describe it with the breathlessness of being there, in that seizure of time that it was.

I am in the Opera House at BAM. I am watching electric, galvanizing movement on stage. It is driven by a band of eight musicians on three levels of upstage scaffolding. Something like techno/metal and unstoppable. A singer stands at the top center, screaming, past language, into the microphone. Light slashes into the murky and fogged environment. The dancers are all young; they all seem possessed.

The dance begins with a group of dancers downstage, arms linked, stopped and still—a body. Not a biological body but a collective, poised for its immersion, its action, in a historical situation. Maybe dancers at a club poised to be taken up in something else, something like a mob scene arranged into the beauty of a historical task, fueled, propelled into it by the unstoppable unhinging of this music and the sweat of the people making it. And they begin.

The body seems to move in and through Jewish folk dance as its central trope, but it is a dance divided, violently, by paroxysms, falls, beseeching, grasping and mis-targeted couplings and jagged un-couplings falling outside of the group, falling from the group, individuated bodies pulling themselves along like carcasses, individual bodies utterly, cellularly, in a state of uncontainable explosion. What I see throughout the piece are the momentary upsurges of the shape of historical traumas, pushing up under and through the skin. The singer, who from the first I think of as the ogre, screams on and on. Sometimes he wears the mask of a gorilla. At times the body dances in front of the band: militaristic, proto-fascistic obedience to the ogre. At the end of the piece, the ogre exhausted, an inscription, a slogan, which has appeared as a fragment before, appears in full across the back screen. It says, "Where there is pressure, there is folk dance." It is a poor, a weak, an ineffectual slogan in light of all that we have seen. It is bafflingly conservative. In the final scene, the dancers resume something like the lives they would go home to after dancing at a club. The thing has, perhaps, collapsed into being only a club after all, a night out. It seems they are unwinding to the saccharine notes of, of all things, Judy Collins singing "Both Sides Now." The dancers are re-individuated. It was a disappointing end; I was drawing a blank about its significance or meaning. But I was nevertheless in an extraordinarily *fizzy* state, a state of excitation, a state of total response. It is hard to tell what had happened in the course of this seventy minutes, this barrage, this untenable degree of physical and emotional extreme, except for the fact that it was an intense time. I wanted to think about why.

I begin with the proposition that the ogre, the ogre/ape, designates chronological time via the evolutionary trajectory that the ape image puts into play. In a small book that is the transcript of his address on the church delivered in Notre Dame Cathedral to the Bishop of Paris, Giorgio Agamben accuses the church of abandoning eschatological time, the time of final things, in favor of a chronological time, which is the time that will not end. The abandonment of eschatology is the church's deferral of the end of its own power. Time without an end is the time heretofore allotted only to hell, the only infinite time. In the parallel secular model of the church's decision, there will be no end to law, only a "hypertrophy of the law,"[1] where the conditions are established for the "truly infernal"[2] perpetuation of power as it is. The human in the ape mask designates the supposedly smooth flow of chronological evolution that justifies the ongoing mastery of the human ruler, the one who has excelled, and who has grafted the traits of the gorilla, as he knows the gorilla, to his own. Hypertrophied strength.

But in the dance we see the Holocaust, we see terrible separations, we see Kafka, we see migration, we see the animals, we see Netanyahu, we see the ruler beast, we see the scuttling of language—the horrifying ogre roar of the experience of all languages collected and demolished—we see refugees. What the dancers dance, what the body does, is an extreme torque, a convulsion in relation to chronological time, of chronological time, even as the body remains in thrall to it, in thrall to the ogre. In the dance, time is lived as hysterical; time is lived as a razor, as an electric shock, as an insect inside. The body is rioting. The body is rioting against the naturalized cruelties of chronological time, against its infernal infinite. The riot gestures toward its end.

In his 2011/12 book that responds to the Arab uprisings, *The Rebirth of History: Times of Riots and Uprisings*, Alain Badiou suggests that many of us live in an intervallic period in which the chronological time of the rulers has once again created a smooth and unbroken seal over history and its riotous, insurrectionary, or revolutionary upheavals. In this naturalized smooth time, revolution is inaccessible, barely remembered. In such a time, says Badiou, "the riot is the guardian of the history of emancipation."[3] The riot, these dancers, this body, are convulsions in relation to intervallic time. It is a break in time, made all the more charged because of the intervallic seduction of the ogre, his wordless violence its own justification and turn-on, the riot's energy driven against the tyrant, racked by his noise, but also the body as a serrated edge against the ogre's breath, he who is the designation of the failure of time to emancipate. In the upsurge, the time of the masters is de-naturalized, brought to the surface, expelled. This time includes the history of the animals—so much of the movement is blended to the movement of animals, animals in distress, animals seeking existence from inexistence. In *Political Mother*, there are two openings to the animals in relation to time: the gorilla man/master, marking the evolutionary sequence, and the riot of the animals against that sequence. Beings and entities who are not human are also subject to the intervallic.

Because of the centrality of Jewish folk dance to this piece, and while keeping in mind that I am thinking about chronological time in general, I want to speak about the folk dance in relation to the chronological time specific to Israel. Shechter's experience in his native Israel, prior to arriving in London in 2002, is, by his own account, pivotal in his work.

Along with his experience training with the Israeli Defense Force, folk dance has been central.[4] Shechter started folk dancing in school in Jerusalem when he was six or seven. He says, "It had everything to do with my getting into dancing."[5] The lines, the repeated and complex interlacing of bodies, the fast circles, the coordinated and fluid group movement, and the steps of the Horah, for instance, are all very much integrated into Shechter's choreography.[6]

The history of Jewish folk dance is a contradictory one. In and of itself, it has historically carried the simultaneity of the injustices of chronological time and the possibility of emancipation into socialism. As Badiou says of the founding of a Zionist state, it was "an extraordinary mixture of revolution and reaction, of emancipation and oppression."[7] We could say that Shechter retrieves the promise of the folk dance, or its excitement, its togetherness, not as an identitarian force, but as a universalizing project of socialism. At the same time, the dancers are stricken by the other side of the Zionist project: what will become the forced expulsion of most of the Arab people already living in Palestine.

The mostly young European immigrants to Palestine—Zionist and socialist, charged with their own historical task, a time within and intervening in chronological historical time, working their agricultural collectives—used European-derived folk dance, especially the Horah, as a core expression of what they wished the new way of living to be. Shechter echoes that history. He speaks of the joy of the folk dance, of its energy, of the togetherness it creates, the feeling of being "all in it together." In her book *Dance and Authenticity in Israel and Palestine: Performing the Nation*, Elke Kaschl writes that the Horah was considered a "dance of affirmation"[8]—joyous, full of life, capable of incorporating new people into the circle. She says, "These dances constituted a means to acquaint newcomers with the ideologies of egalitarianism and communalism determining life in these collectives. . . . The dancing circle of the Horah, where 'no one is first and no one is last' was turned into a romanticized symbol of the socialist community of togetherness and pursuing a common goal."[9]

But the situation of the folk dance began to alter in nature during the Second World War, to become a foundation for the chronological time of what would soon be the Israeli state, which was established in 1948. Choreographers on the kibbutzes began rejecting the European roots of the Horah. They wanted a new identity for the children born in Israel, Sabra, an identity that would be specifically tied to the new land, "the beautiful homeland Israel,"[10] which in turn was linked to the supposed ancient and Biblical past of the Jews in Palestine. And so choreographers decided to literally invent a Jewish folk dance, and they did so, astonishingly, by appropriating the Palestinian folk dance, Dabkah, which eventually became the Israeli Debkah.

In 1947, Rivka Sturman choreographed *Debkah Gilboa*. She created it during a clash over land between her kibbutz and Palestinians at the mountain of Gilboa, in which the kibbutz settlers were victorious and a new settlement was created on the top of the mountain. The dance was based on a Biblical story—in this case the battle at Gilboa, which the Israelites lost to the Philistines. Kaschl says of this dance, "To express Zionist victory through dance, Sturman turned to the Arab Dabkah, the dance style of the enemies, as a model of artistic inspiration. Reinventing Dabkah as a *Debkah Gilboa*, Sturman thus not only celebrated a one time military victory, but staged the general defeat and ultimate erasure of the Palestinian presence around the area of the Gilboa Mountain."[11]

Like *Debkah Gilboa*, the first folk dance created to celebrate the establishment of the state of Israel was performed in the style of the people it was erasing. Performed on the first Independence Day in 1948 and called "So Let All Thine Enemies Perish," it was a mass spectacle in which, in Kaschl's description, "hundreds of soldiers from various army bases came together, performing the dance in two rows opposite each other, mimicking acts of attack and final triumphant victory."[12] Large public folk dance events are still used to commemorate Independence Day, still actively maintaining the concoction of eternal, linear, historical time-telling to certify the belonging of one set of people in Israel at the expense of others.[13]

At the heart of Jewish folk dance is the word "davka," which means defiance. The folk dance is meant to express the hope and the pride of the Israeli people and their ability to overcome all struggles and difficulties. The call of "Davka!" means that "in spite of the situation, he danced."[14] Shechter's ambiguous phrase "When there is pressure, there is folk dance" is clearer when put in relation to the Davka slogan he must certainly know, in relation to "in spite of the situation, he danced," but also deeply complicates its identitarian chronological assertion.

Through his simultaneous presentation of folk dance and the distortion of it, a break in chronological time is implicit in the piece. Jewish folk dance bears within itself a remote socialist past, its own break in time, and the *invention* of a chronological time which erases the fragment of revolutionary time. Through it we can see chronological time being made up, and specifically made up also with a view to establishing itself as natural and eternal. We see the lie of its longevity—its arc into the future and out of the past. We can catch a glimpse of the erasure (in this case of a people) on which that unalterable chronological time assumed to be the mechanism of history, what I will call historical processionism, founds itself. History, and the form of time assumed to be driving it, integral to it, founds itself on injustice.

And then there were the suspensions of the dance, which were to me the most arresting feature of all. They were both simple and very full, with the dancers holding, standing almost limply, with arms loosely raised straight above their lowered heads. These suspensions were a powerful *cessation*. As such, they were an experiment in ending, in trying out the end. They were disconnected even from the time of the riot. To return to Agamben's language, they are a "disconnection" that is necessary so that we can "grasp time."[15] I can use another phrase of Agamben's to say that the suspensions opened "the time that remains, the time we need to end time, to confront our customary image of time and to liberate ourselves from it."[16] The suspensions looked to me like the conditions of possibility for the end of the time that legitimates its own infernal/eternal rule. They are a cessation through which we both grasp and sever our place in chronological time.

And so, why the elusive title of *Political Mother*? In the first place, we can take mother as the point of origin, as designated by the church and state—origin, that is, assigned gender and gender attributes, including passivity; the pink point of origin from which strings and chains of lineages and repetitions unfold for all time. But perhaps the mother in a time of riots confiscates those designations, and becomes a political mother. A political mother without gender, without chronology, but still accorded the principle of birth and generation, a mother who reorganizes the conditions of generation, the possibilities of generation, without obedience to chronology, seizing from the state its (as Badiou says) "monopoly on the definition of political time."[17]

And yet, in the end, the dance, considered as a riot, is only what Badiou calls an "immediate riot." In the historical riot, the possibility of duration and new worlds exist. The immediate riot, a "tumultuous assembly of the young,"[18] remains, unlike a historical riot, unextended, and the type of subjectivity it creates is indistinct. There is no unifying strong slogan. There is no affirmative demand. In *Political Mother*, there is instead the soothing reassurance of a song written, ironically, in another time of riots, the 1960s, now forgotten. And there is only the weak slogan.

But we might think of the immediate riot of the body onstage, warping the time of the ogre however temporarily, as an experience of what Badiou calls the "*latent* riotous subjectivity of the intervallic period."[19] Watching it is to be privy to the intense time of the riot, a time when people are acting on a force of energy that is not supportable in everyday life, and which will inevitably have to dissipate, but which nevertheless *happened*. *Political Mother* offers an intense time in the theatre, in parallel to the intense times of the riots in the world.

Fire

The force of breaking time, of insurgent time, of creating a time that will bring an end to the infinity of power, all of that in *Political Mother*, began to occur to me metaphorically (and sometimes literally) as fire. The image began to shape my thinking for the book, and for its title. I wanted to insure that the idea of initiation was compounded with the conflagrating force that fire has often had in insurrectionary situations.

The temporal initiations in this book, the initiations of new structures and determinates for history, are mostly not accompanied by actual fire, with the exception of Chapter 2, on plays on the Haitian revolution, and a performance by Cassils, discussed in Chapter 4. I mean for the idea of fire to assert itself throughout the text in the following way. Fire is in many ways relied upon, both actually and metaphorically, to signal the interruption of history as processionism. Fire denaturalizes that history, sometimes creating new forms bearing in their arrival the signs, truths, and energies of pasts that came before. During the time this book was developing, for instance, the movement against police killings—especially of young black men—burst with fiery force into the fabric of life in the United States; the movement as contemporary innovation carried and continues to carry traces of forms of historical black radicalism. It is, for a person living in the United States and creating a book between 2012 and 2017, the signature form of fiery initiation.

The photographs of the Ferguson, Missouri, riots of 2014 are classic anywhere-and-everywhere photographs of the riot situation. The figures in the photographs are in extreme motion, stopped by the camera in emblematic gestures of triumph, defiance, and rage. One of the finest and most iconic is a woman who is standing in the extreme foreground of the picture with her back to us.[20] It is night. The time of the riot is also the night, when the fire is most visible in its fury and force, making people visible through the force of their gestures in the clear cut of the silhouette. This woman too is in silhouette. She is visible from just below her hips to her head, and her body is centered at the near end of what appears to be an illuminated golden path, receding with perfect perspective into the fire-induced smoke of the near distance. The only marker of structure clearly visible in the smoke that fills the image, rising from ground to sky, is the "golden arch" of a McDonald's off to the side and some ways down the road. It glows and glares its hot yellow like the heart of all that is most malevolent. To her left, the smoke glows blue. The woman has her hands raised high and outward on either side of her head. With each hand she is "giving the finger" to . . . whom? There is no other person in the picture. She faces the golden path as if she is about to embark, moving toward something new, walking into or toward the fire, the apparent center of the action, or into the shields and guns of police, armed with the strongest emblematic expression she has. What is that cleared, golden, fiery space she faces, her angel's wings inescapably outspread with the winds of the future—but perhaps, unlike Benjamin's angel of history, holding strong against those winds while she faces the resurgence of the history of the riot, from the wreckage of its pasts, maybe not only wishing but actually "awaken[ing] the dead."[21] There is something in her, too, as in Benjamin, of "the prophetic gaze that catches fire from the summit of the past,"[22] of the "the explosive materials" that are "latent" in the past that can be ignited.[23]

I think it is safe to say that every living being fears fire. Its power to devour and destroy buildings, forests and all their entities, animals, possessions, human lives, is terrifying. Those who have been close to it describe it as thunderously loud. It takes us unaware. Its blackened landscapes can sear the heart. It is, of course, hell. And yet, perhaps because of all this, it is a powerful weapon, in its capacity to strike, to heat, to ignite, to light stretches of the dark with its beauty. It can be used in the hands of fascism, of terror, of brutality—by the forces of brute reaction. But it also can be, and has been, both an actual and metaphorical force against processional history and the injustice it lays down in its tracks. It is deeply mythic, and politically accessible. Peter Linebaugh and Marcus Rediker, writing about *The Many-Headed Hydra*—the insurrectionary eighteenth- and nineteenth-century collaboration of workers, slaves, and sailors joined by work on the Atlantic and at its edges—say that fire "was the most accessible of weapons among the dispossessed,"[24] and that "it remained a weapon of liberation. If it threatened apocalypse, a new world might yet arise from the ashes."[25] Fire appears over and over in this sense, in literature, poetry, music, political theory, political tracts, and military strategy. It is everywhere in the political situation in which I write. It is in Kendrick Lamar's electrifying 2016 Grammy performance of "The Blacker the Berry" in the form of a huge street bonfire. It is in Beyonce's 2016 visual album, *Lemonade*, as the street behind her roils into flame.

Fire also clears the ground: riotous fires, guerilla warfare fires, accidental fires, forest fires. Something can arrive, something can be made, vital new growth can start, in the cleared ground. Rustom Bharucha writes about this in *Terror and Performance*. He criticizes a "rhetoric of the theatre" around fire that infuses itself with Artaud's signals through the flames and the sense that a performance rises like the phoenix from the ashes of the previous night's performance. This rhetoric, he says, is deployed to speak about the alleged "transformative power of performance."[26] In what I take as a significant move that works from the literal consequences of fire, he replaces this rhetoric with fire as the razing of what exists, what is in place. It leaves behind "a vast emptiness," the grounds "not just for the lesson of renewal [as with the phoenix] but of radical hope."[27] Two remarkable performances stage the razing to the ground, the emptiness left behind, as grounds for radical hope, or, in my language, for an initiation of new determinates of history, new times within and of it: Brandon Jacob-Jenkins's *The Octoroon* and The TEAM's *Architecting*.

The Octoroon, which I saw in fall 2014 at the Theatre for a New Audience in Brooklyn, New York, begins with the melodramatist Dion Boucicault bemoaning the burning down of his own Winter Garden Theatre in New York in 1854 and his reduction of fame and fortune as a consequence. The play in which he finds himself as a result, Jacobs-Jenkins's remake of his own, torques the time of slavery, the time of the plantation, the time of minstrelsy. It ends in the fiery offstage conflagration of the steamship. The theatre is completely cleared. After a moment the actors drift back onto a darkened space, emptied, the space left behind by the fire, clearly both the space of the theatre and of a world. They sing an intensely moving song, barely visible, lined up laterally facing us across the lip of the stage. They sing exactly about the opportunity to imagine how we can fill worlds now, after a seizure of processional history, of the time of slavery: "When you burn it down there is a clearing / When you burn it down there is no trace / What do you put there in its place?"[28]

In The TEAM's *Architecting*, the entire space of the performance is also cleared, physically, after the centerpiece of the architect's post-Katrina gated community—a Ninth Ward home left behind by a fleeing black family, now marketed as a nostalgic remnant—is burned to the ground. Franklin Delmore McKinley, whose home it was, returns from his exile in Arkansas leading an army of people in "varying degrees of dark, an army of black people,"[29] to burn it down. This play is full of fire and what gets done in its clearings—there is Tara, from *Gone with the Wind*, there is Atlanta, there is the Chartres Cathedral, rebuilt by thousands of anonymous people. The play makes a visionary space, with Tara, *Gone with the Wind*, plantation Atlanta, neoliberal development, gone up in flames. In the clearing there is "open space . . . a dynamic theory of history,"[30] a space and a time for "an affirmative collaboration,"[31] a new form of architecting.

The Octoroon and *Architecting* exemplify the striking aesthetic, grounded in innumerable actual conflagrations of theatres, of imagining theatre burning itself down, burning out its own stage, in an act of specifically temporal defiance. Here it takes unto itself the function of people burning their own neighborhood. In the course of the riots that broke out in response to verdict after verdict exonerating policemen from guilt in the overwhelming number of instances of police shootings, neighborhoods in which black people lived, where the murdered were from, were set on fire. We are all familiar with the commentaries of baffled pundits and the interview comments of "the man on the street" who cannot see the point of rioters "burning down their own neighborhoods." (This is a phrase that is so common that it does not even bear citation.) Black Panther Eldridge Cleaver said "community liberation" (burning) was "to free our people locked up as they are in Urban Dungeons."[32] Haiti, burned by the slave revolutionaries, was not, of course, an urban dungeon. But the ghetto is a colonized space, created as an irredeemable spatial difference, and also as an irredeemable temporal distance. Those in the ghetto are set apart from the chrono-normative time of the others—they do not keep up, they fall behind. So, the fire of these particular contemporary riots, both like and unlike the fires of the slave revolts, marks the burning down of the space of non-liberty, of the time of imprisonment—in order to hold open a new present, a time of the emancipation. A song from the 1970s by the Watts Prophets, "Dem Niggers Ain't Playing," goes like this: "Look at those flames lightin up the sky / Ain't never seen fire shootin up so high, (repeat) . . . . Ever since they passed them civil rights those fires have been lighting up the nights / and this city isn't going to stop until we all have equal rights / looks to me like them niggers ain't playing."[33] What these theatrical and actual clearings have in common is the space of a beginning, an inception, and the open question of what will, now, be initiated. It is the open question of what, now, differently, will claim the name of history.

Theatre and Beginnings

This sense of the theatre, performance, as a clearing, an opening, a beginning, stands in contrast to a more standard dramaturgical, structural, philosophical, and aesthetic emphasis on theatre's closures, its means of bringing its content and structure to a necessary end. My intent is a shift of emphasis from endings to beginnings, to openings. For instance, Beckett, canonical and yet still our contemporary, one of the great innovators of temporality in the theatre, among other things, gives us in *Endgame*'s Hamm a character who knows something very well. It is that at the beginning of all plays, the audiences are waiting for us to *begin*, to set something in motion. Hamm sets out to engage us with beginnings, waking, removing the stauncher from his face over and over, to begin to play. This is true whether we interpret the play as a cycle that is repeated every day down to the last detail, including the exhaustion of resources that are tallied, or whether we think we are seeing the approach of the endgame itself. The play is saturated with beginnings. In neither *Endgame* nor *Waiting for Godot* are the characters at the mercy of the chronology that seems to define their time on stage, the time of their days. This is because there is no verifiable indication of an end of any kind, that there has ever been an end—of a day, of an action, of a sequence, or of a life.

In *Waiting for Godot*, Didi and Gogo's one gambit toward even thinking about death, hanging themselves in the tree, is quickly recognized by both as conceptually and materially impossible. Somehow, they escape the assumed flow of time by which the lives of others (we the audience and everyone else in our world) pass from beginning to end. Somehow, despite their perceived helplessness and paralysis, they have the power to begin time, to initiate time, to repeal the law of the end (however much they might desire it as the end of their miseries and however many times they announce it).

One of the great Beckett critics, Martin Esslin, remarks that *Waiting for Godot* foregrounds "time itself" since it is about waiting, and it is when we are simply waiting that we become most aware of the tick tock passing of time, "the flow of time in its purest, most evident form."[34] I would argue, on the contrary, that Didi and Gogo subordinate waiting to beginning. That is, they play with time, they see what they can do with it, they initiate their own sequences. We laugh, but not only because their situation is so futile, their own awareness of their hopeless situation so caustic. On YouTube, there is a one-minute-and-thirty-nine second promotional video of *Waiting for Godot* with Ian McKellen, Patrick Stewart, Billy Crudup, and Shuler Hensley. As Didi and Gogo, McKellen and Stewart clearly delight in each other's masterful skills as actors and, it seems, in each other as two exquisitely close, aging human beings. The video begins with the little Act II sequence in which Didi prods Gogo to say he is happy.[35] Once they agree that each is happy they say, "What do we do now, now that we are happy?" Of course, the question is a fantastically funny evacuation of the concept of happiness itself. But in addition to this, the two actors play and invent, frolic and dance, and entertain each other, and laugh, and beyond even their ability to help themselves take pleasure in what they do with time. The video editors restore us to the quieted characters and the line, "Well, that passed the time," with its response, "It would have passed anyway." But we have witnessed an internal bubble in chronological time—"it would have passed anyway"—which has been full of inventive play, games, with each game played as if it were a new beginning.

In all plays or performances, every actor who walks out on stage begins something. We trust actors and performers to walk out there and begin *something*. We are (overly) familiar with the idea that performance is a rehearsed repetition. And of course it is. But there is always something new that happens, some crack in what has happened before. To begin a performance is truly to begin something in which something unforeseen and unknown will take place or emerge. In a sense, there are small clearings internal to every performance, clearings for beginnings.

Initiations of History

I have been using the word "initiate" in my sentences in the course of these last pages. I have made it nearly synonymous with "beginning," but there is also more to it. Initiate, or initiation, carries here the implication of not only a beginning, but the subsequent shaping of that time which installs into processional history a different kind of structure or set of determinates. Here is the definition of initiation upon which the material in this book depends.

> To initiate is to create a disorder by means of a temporal innovation within processional history. To initiate is, by means of that disorder, to nominate an alternative possibility for what it means to live for those whom "history" has meant to vanquish, and for those who have been consigned to merely spectating its supposedly unalterable flow as they move toward death. Even if it is tentative, or small, it is a difference that unbinds the temporal authority of power. To initiate is to create an askew trajectory that ignites its encounter with the straight

line of processional history with the illumination of newly appeared capacities.

To initiate is to do what Ni'Ja Whitson did in *A Meditation on Tongues* at the American Realness festival at Abrons Art Center in New York City in January 2017. Whitson, along with Kirsten Flores-Davis, is dancing in or as dead queer black men, here again, in our presence. Flores-Davis begins the piece naked, tall, hair buzzed, statuesque, challenging us with her eyes as she moves through us as we are crowded outside the performance space. She is followed by Whitson, in baggy street clothes, a lover who stands up against Flores-Davis in the hallway into which they move, seductive, hot, while Flores-Davis dresses. Then Whitson guides us into and through the hallway, lined with memorial sites to queer black men who have died. We hear a tape of a man who is assaulted verbally on a bus for his queerness, called a "bitch," who responds by calling back to the assaulter something like "the real bitch is waiting for you at home," implying the assaulter's wife or girlfriend. It is a quick moment of paralyzing consciousness, this exchange of one violence for another, one against queer, the other queer against women. This contradiction is not so much resolved as dissolved into a certain kind of activating force in the bodies of the performers, who—in a sometimes-cruel choreography of hard intersections with each other, bodies sometimes tender toward one another but also alienated and combustible—work their way into the pain of the queer black men, move ever closer to and into it.

They are working from history, retrieving—the corridor of the murdered through which we have passed, and clips from the film *Tongues Untied* by Marlon T. Riggs, which documents the historical resistance of black gay men, projected at the end of the piece. But they are creating a disorder. They are making a form of life, or capacity for life, that is out of the ordering of the history that vanquishes the experience of gay black men, but also even out of the ordering of that counter-history. The counter-history is the support, but not the content, for the initiation of a new form of body, of experience, of politics, a new assembly of political queerness that outdoes the misogyny from the queer black man, a fashioning, through pain, at my feet, on my feet, of a potentially different political praxis or capacity.

I am seated centrally in the Underground Theatre, on the top platform, just at the opening used for the audience to climb up on the platforms to the right and the left. Before me are two broad openings in the stacked platforms, forming a sort of steps. In a grueling choreography, Flores-Davis dances spread out on those steps, coiling and uncoiling, in spasm, in unutterable grief and desperation, not protecting her own body in any way, reaching out, vulnerable. It seems to me that she is dancing a death by poison, an interior so ruined, death as a poisoned condition of being for US-African people, history as poison. It is all I can do not to reach out and put my arms around her. I wonder what would have happened if I had. When she finishes, she puts her head on the lap of a spectator to my left, down one step. She rests, waits. The spectator does not move, does not rest her hand on the dancer's head. Whitson moves to the steps. She sits and Flores-Davis moves to face her, their bodies, their tongues, their faces close up, their sweat, their sexuality, their desire, their life. They are focused in an act of the illumination of a capacity for life that finds a temporality for itself that does not rely on the processional history that has tried to vanquish this richness, this desire. Each pulls a small three-by-three-inch piece of paper from somewhere in their clothing. They paste it up against the sweating chest of the other. They read breathlessly to each other, their urgency flowing over the divisions between the lines—a love poem, a meditation on tongues, by Essex Hemphill, the poet who was a guiding light for the black gay community in the 1980s and 1990s and who died of complications from AIDS in 1995. When the performance is over, I see that one of the pieces of paper has fallen to the floor in front of me. I have it, crumpled and a little bit torn by their heat and urgency, and this is how it appears. I have excerpted it.

times are lean pretty baby
the beans burn to the bottom of the battered pot
let's make fierce love on the overstuffed hand-me-down sofa
we can burn it up too

our hunger will evaporate like money
I can smell your lust
not the pot burnt black with tonight's meager meal
so we can't buy flowers for our table
our kisses are petals
our tongues caress the bloom
who dares to tell us we are poor and powerless
we keep treasure any king would count as dear[36]

There is, in this performance, an initiation of history. Here, in this book, history is

> the manifestation of a radical human capacity, deployed along with or in the interest of other entities, and found in temporal interstices and innovations, in order to inaugurate universal (belonging to everyone and excluding no one) conditions of justice, recognition, and equal distribution in particular local and/or globally related situations.

Linked to initiation, it is the grounds for criticism of extant conceptions of history and temporality (in theatre and performance and in general), and the foundation for thinking toward changing time.

Processional History, Historicism, and Walter Benjamin's Radical Time

My thought for this book would not have been possible without Walter Benjamin and his theoretical imagination, as it devoted itself to a critique of what I am calling processional history, and to a suggestion for a time that replaces it.

The historical assumptions against which Benjamin launches his fiery innovations are linked to a temporality that rests on what is referred to variously as cosmic, objective, or Aristotelian time. Before I move on to a fuller discussion of Benjamin, it is necessary to say a little about objective time (to which I will return in the section on phenomenology in this chapter) and its relation to what is called history.

Simply put, Aristotelian time is a succession of "instants." The instant is both the hinge between what has happened and what will happen *and* the separation between them. Time is thus a measure of the movement between what comes before and what comes after. As Agamben phrases it, time is a "precise, infinite, quantified continuum,"[37] or, as Peter Osborne says, Aristotle's time is "an objective time of serial succession."[38] Originally perceived to be of the cosmos, it remains cosmic, or objective, in the sense that forms of subjective time or human time (introduced later, for example by St. Augustine) do not influence it. It is upon this idea of unalterable "serial succession" that the history under critique here builds itself. Clearly it is refreshed and invigorated, vastly complicated, and variously shaped by the demands placed upon it by changing social, political, and economic formations. And yet, the way in which it undergirds dominant conceptions of history remains constant, reliable even, as its force and influence are rarely acknowledged when we speak of history. That is, temporal assumptions and practices remain the hidden content of history, implicit but unexamined. If temporality is articulated in relation to history, it is as an inseparable substructure that motors us along the path of serial succession. It seems to exclude intervention, or else to be capable of restoring any intervention that happens back into its unalterable flow.

Because this notion of temporality and this notion of history service each other so well, it is both difficult and urgent to disentangle them. Disentangling them makes it possible to see and articulate the varied formations and alterations by means of which temporality is actively and continually constructed for ideological and material purposes. And, most importantly, it becomes possible to release temporality from the exoskeleton that determines its use so that it becomes available in other forms. We can identify our task within the following well-known description from Agamben of the imbrication of time and history:

> Every conception of history is invariably accompanied by a certain experience of time which is implicit in it, conditions it, and thereby has to be elucidated. Similarly, every culture is first and foremost a

> particular experience of time, and no new culture is possible without an alteration in this experience. The original task of a genuine revolution, therefore, is never merely to "change the world," but also—and above all—to change time.[39]

For Benjamin, history must be concerned with our past and its retrieval. The kind of retrieval in which he is interested specifically negates the formula of what he calls historicism with altogether different temporal propositions. Historicism is a point of view, an "indolence of the heart,"[40] which empathizes with the victor in each situation in which the historian immerses himself. History practised in this way thus results in the linking of a chain of victors, since "all rulers are the heirs of prior conquerors." History is a succession of rulers, victors, stepping "over those who are lying prostrate" and carrying the spoils won from their conquests in "procession."[41] Historicism's lulling "once upon a time"[42] is a series of one after another, transitions from one to another across the present, the Aristotelian hinge, which disappears into the future almost as soon as it appears, which keeps history moving forward. Historicism's content, used to support the procession of the victors, is an "additive . . . mass of data."[43]

Benjamin's most famous passages on temporality and history are included in his 1940 essay "On the Concept of History" and in "Paralipomena to 'On the Concept of History,'" from which I have been quoting. In the state of emergency (which is always the rule for the oppressed) that is fascism in the context of which he is writing, he demands an altogether different conception of history and, therefore, of time. In particular, and again so importantly for my own project here, Benjamin's thought separates decisively from the assumption that time and history are directed toward the future. His writing here is directed against the Social Democrats. They succumbed to a processional view of history such that "nothing has so corrupted the German working class as the notion that it was moving with the current," participating in "the illusion that the factory work ostensibly furthering technological progress constituted a political achievement." They championed a renewal of the work ethic and a valorized conception of work that was "tantamount to an exploitation of nature"[44] in a way that did not recognize the inter-exploitation of the working class and the natural world. By contrast, Benjamin presents what he says is a temporal image in which, as in socialist utopian visions, labor would help nature "give birth to creations that now lie dormant in her womb."[45] Benjamin here provides a novel (and presciently environmental) conception of labor and the working class not as forward-directed, tearing ahead by devouring raw natural materials, but as working from what is immanently possible, available from an immanent interior of what already is.

Further, misunderstanding the source of the working class's energy for uprising, its "hatred and its spirit of sacrifice," the Social Democrats wished to mold the working class as "the redeemer of *future* generations." Benjamin says, famously, that the working class's hatred and spirit of sacrifice are in fact "nourished by the image of enslaved ancestors rather than by the ideal of liberated grandchildren."[46] Here, the historical materialist, as opposed to the historicist, offers a relation to the past and a retrieval of images and ideals from the lives of the oppressed, or the "anonymous,"[47] instead of to the future. This is the hinge on which Benjamin turns temporality and history, like the heliotropic action he attributes to flowers, toward aspects of the class struggle, things very different from the spoils of the victors as they march on toward their future victories. These aspects are "confidence, courage, humor, cunning, and fortitude," which "have effects that reach far back into the past" and thus "constantly call into question every victory, past and present, of the rulers."[48]

"Empty, homogeneous time"[49] is the famous description given by Benjamin of the time that undergirds nineteenth-century notions of progress, or what Rolf Tiedemann calls "positivistic historicism,"[50] especially those espoused by the Social Democrats; progress is "inevitable" and "boundless in keeping with the infinite perfectibility of humanity."[51] Benjamin calls the history driven by this time a "continuum of history." What the "revolutionary classes" do is to "explode" that continuum. The explosion makes history a "construction whose site" is "time filled full by now time." Pasts become "charged with now-time" when they are "blasted out" of the historicism's continuum.[52]

What is crucial for Benjamin is to *still* that progression, to stop it—to contest in a revolutionary way the historicist's presentation of history as a series of presents which are only transitions toward the future of the next ruler. He ascribes this action to *thought*, the process of thinking about history that the historical materialist, as opposed to the historicist, does. In addition to moving, traveling (presumably in history) thought must bring itself to a standstill. The thinker stops, so that they see and feel the full tension of the "constellation" of things around them. Their consideration of the historical object must take that object as a "monad" or a full, self-enclosed "image of the world."[53] That is, not a world in process, moving forward, but a world present in its entirety here in this standstill. Benjamin calls this "arrest" of happening "messianic"; in this instance, he seems to mean by messianic that the cessation of happening is so important, so vital, that it is the only way by which revolutionary redemption of the past can occur.

Benjamin speaks of the Jewish teaching in which Jews were not to think of the future, to only look back to remembrance of the past, even as any moment might be the "gateway in time through which the Messiah might enter."[54] Thus, looking back is the correct orientation while leaving an absolute openness in time for the expected redeemer to make his appearance. It is not to wait for the Messiah, looking into the future. Put differently, "in reality, there is not a moment that would not carry with it *its* revolutionary chance."[55] The (messianic) arrest of happening "blast[s] a specific era out of the homogeneous course of history."[56] The historical materialist, at a standstill, is in both their own era and an earlier one, forming a constellation. What is thought of as the present (that hinge into the future, according to Aristotelian and historicist time) becomes now-time, a constellation of the historical materialist's own time and an earlier time in which revolutionary possibility—revolutionary interruption of the continuum, like the messiah's interruption of time and the release of revolutionary energies—is retrieved. As Agamben says, "The time of the messiah cannot designate a chronological period or duration but, instead, must represent nothing less than a qualitative change in how time is experienced."[57]

The materialist historian, at a standstill, grasps an "image of the past that is not recognized by the present as one of its own concerns," traces of utopia from "authentic historical time"[58] that awaken the historian with a flash of recognition in a now comprised of this awakening. Their practice of history, as Tiedemann says, is "bound to the practical liberation of humanity."[59] Benjamin calls this image the dialectical image.[60] If historical dialectics depend upon a theory of forward movement, the dialectical image and dialectics at a standstill hold the then and the now together, such that the past becomes now, recognizable as now. With the dialectical image, Benjamin replaces the dialectics of Marx, in which revolution is awaited through long phases. In Benjamin, the messianic entry is the end of history itself, where "the classless society is not the ultimate goal of progress in history but its rupture, so often attempted and finally brought about."[61] Agamben argues that Benjamin understood St. Paul's message well: the messiah is not the one "who comes," in the sense of will come, but is instead "he who never ceases to come."[62]

The implication here is not only of a refusal of the time of the future, along with the chronology of historicist time, but the idea of the possibility of repeated messianic cessations of the flow of time and the redemption of time that has been, dreams in the past of the arrival of what capitalism can never deliver. We (the collective) are as if awakened from them. We carry out "what has been" when we remember the dreams upon awakening. Benjamin writes, "The compelling—the drastic—experience, which refutes everything 'gradual' about becoming and shows all seeming 'development' to be dialectical reversal, eminently and thoroughly composed, is the awakening from dream."[63] At one point in Convolute K of *The Arcades Project*, Benjamin speaks of the past that saturates the now-time of this awakening as more acute, more real, more concentrated, than it was in its original occurrence: "This dialectical penetration and actualization of former contexts puts the truth of all present action to the test. Or rather, it serves to ignite the explosive materials that are latent in what has been."[64] It is, in fact, this statement and its implications that resonate throughout the book, or belong to its substructure. What is genuinely historical (rather than archaic, in the sense of historicism's consignment of things to the un-retrievable past), "authentic historical time, the time of truth," can only be an image in which what has been, a kind of dream of revolutionary fulfillment, becomes recognizable or legible in a now. The conventional relationship of the present and the past is here overcome, replaced by a "relation [between] what-has-been [and] a now," a dialectic which does not move forward but is at a standstill, in which "truth is filled to the bursting point with time."[65]

The Now, the New, and the Present

The now, the new, and the present are categories of time which need some differentiation for the purposes of this book. Benjamin, and his extraordinary contribution to intervening in processional history, helps here too. According to Peter Osborne, Benjamin's temporal/historical theory is sourced in a crisis of memory, remembrance, and tradition. For Benjamin, tradition (or the lived memory of the experience of the oppressed) is erased during the nineteenth century by the scientific positivism in knowledge production that began to replace experience. The thick, tangled relationship between memory and lived experience in the unprovable past is invalidated, replaced with a form of knowledge that has verified certain events as true and produced "an abstract continuity with the past in a naturalized and merely chronological form."[66] This replacement became a structural component of the historicist notion of progress. In historicism, interruptions in history (blasts of the historical continuum, revolutionary experience accessible perhaps in memory) lose their vibrant singularity to become merely instances of "the new," an abstracted temporality that is then applied to history as a whole. Historicism becomes modernity's structural replacement for the kind of historical time established by tradition. Historical events are conceived as a kind of mass production of the new, making all events the same, and this projection regularizes and regulates disruption. So traditional memory, unique experiences of the past, are regulated through subordination to historicism's leveling agency. They become abstracted enduring values (in the form of "heritage"), and/or productions of the new moment in an incessant chain of the production of the new.

This frame for the "new" is important because I use the word repeatedly in Badiou's phrase "the new present," even as it is a temporal category that, in regular usage, does not have a place in this book. The new is different from the "now" of Benjamin. As Badiou elucidates at length in *The Century*, the new is the great catalyst for the twentieth century, for its modernism. The impulse of the new is precisely to sever itself from historical precursors. The brilliance of avant-garde art and thought seems a strange bedfellow of historicism, particularly as the moment of severance from the old is violent and ruthless. In the avant-garde, the new was a question of commencement, of beginnings, on the ground of the total destruction of the old, a historiography of the commencement, as opposed to historicism's chain of events. For me, the word "commencement" is important and is a concept to separate from "the new." It implies, to me, the sense that is important in this book, and not unlike the arrival at any moment of the messiah, of a beginning not awaited, but initiated. The emphasis is on beginning, rather than the separation implied by "the new": a separation which is embedded in a historicist notion of history. And yet the notion of the new always walked a fine line, because just over the edge from commencement it belonged to processional history. The new, of course, also implies incessant forward movement for the sake of the future and an absorption of commencement into abstract and uniform units of temporality that occur, according to processional history, regularly over historical time. And it is in this sense, of course, that the new became a central motif in commodity capitalism increasingly throughout the twentieth century as corporations learned planned obsolescence and moved forward toward that future in which many of us now live—of waiting with bated breath for the newest Apple product. The new is only the next product birthed by the future, which is continuously moving on, followed by a populace afraid of getting left behind in the past. The innovation of avant-garde commencement yields to the lull of historical instants strung together, in the promise of the continually regenerated and meaningless new, toward the future.

Badiou's concept of the new present, as elaborated in Chapter 2, has much more to do with Benjamin's now than with the twentieth and twenty-first centuries' new. For Badiou, as for Benjamin, the new present is new because it is inflamed with the emancipatory energies, or an Idea (which must be for and include everyone, a non-humanist radical universal), of a present that existed in the past. Not unlike a dialectical image, but stripped of Benjamin's sensorial, image-based fusion of theology and Marxism by the not less passionate world of the mathematical formula, Badiou's new present is one in which an Idea is inflamed *again*, even if it has been covered over, occulted, for years, centuries. That is, the word "new" as used by Badiou carries a similar implication of the new constituted by the revolutionary energies of earlier presents. It is, more than or in addition to a commencement or a beginning, an initiation.

It is Badiou's theory in particular that has guided me into the confrontation with an ideology of death, especially in Chapter 3. The new present irradiates the biological (bio-political) consignment of the body to death, the feeling of plodding unidirectional existence, with a refusal of finitude. It replaces it with the possible practice of living in the flash of that present which affirms that what is emancipatory exists again, over and over, a "continuous creation."[67] It is to live in a present that may be dense with radical pasts, a present that is created with them, and to "incorporate oneself" into that present. To move into, to be of, to practise the existence of the beauty of past presents in the new conditions of the present is to practice "eternity" as opposed to death—eternity that has been created, and not some transcendent eternity. In attaching ourselves to these presents, we attach our existence to, and as, these "fragments of eternity."[68] We will see a similar concept in Antonio Negri's work in Chapter 4. The eternal is the name given to radical pasts that keep moving back into existence in shapes defined by the new situation, to radical pasts that will not go away, that keep breaking apart the sutured moment to moments of smoothly passing chronological time/processional history. To adhere to them is to become to a certain extent immortal.

We will see in Chapter 2 how the idea of the eternal is also used in the sense of the (transcendent) way in which power structures, the state, the church, capital, figure themselves as without end, even as the interior of this structure is death: the processions of wreckage, genocides, immiserations of all kinds. But in Badiou, and this is key for me, to live in the new present and its complex illumination in new worlds of eternals is to live in "affirmative joy,"[69] rather than in subordination to death, as the frame by which life is measured. The new present offers this.

The Field: In Three Sections

1. The Present and Presentifying in Contemporary Temporal Theory

The present as it has been conceived in contemporary temporal theory, particularly in the arts and especially in phenomenological work on time, is a very different thing from Badiou's present, although it is also important as a certain kind of refutation of chronological time. This conception of the present is not persuasive to me as a critical apparatus, perhaps because it is so closely associated with what we might consider forms of capitalist time, even as it hopes to disrupt those forms. It is also important to say that, although "the present" and "now" are often understood synonymously in everyday language, I wish to keep them disjoined in order to preserve the meaning of Benjamin's now and Badiou's present.

Christine Ross's book on the "temporal turn" in art is an example of an attempt to conceive the present as an intervention in processional future-driven history. She describes what she sees as the contemporary artistic move to suspend forwardness to remake the relations between temporality and what she calls historicity or "the condition of becoming historical."[70] What artists working in temporality do, she says, is to *presentify* the modern "regime of historicity,"[71] by which she means complicating the relations between past, present, and future. She proposes that "once the hollow idea of progress has been hollowed out of its content," artists are "driven by a concern for the future *in light of* the present and the recent past" rather than providing "a new content to the future."[72] The future, she says, is "removed from its modern role" as the "initiator of change." The present, instead, becomes a malleable negotiation of past and future. The present "gains in texture, thickness, influence and complexity."[73]

The temporal turn changes the idea of "the becoming present of future events"—which conceptualizes the future as that into which the present disappears, or the present as always the site of the realization of the future—into "the becoming present of past events, and then their becoming future,"[74] which seems to establish historicity, or becoming historical, as the central instance of the continuum of time. The myriad ways in which humans, she says, experience time, its phenomenological effects, its passing, constitute the lived durational time that contrasts to objective time. Historical time, she argues, differs from objective time (which I have also been referring to as cosmic time). That is, she claims for phenomenologically experienced time (which she calls temporality, as differentiated from objective time) the entry point for humans into historicity. It is the experience, in this malleable present, of this "human possibility to phenomenologically manipulate temporality,"[75] that allows humans to complicate time and establish their historicity.

For Ross, it is important to keep in play both objective time and complex "human" phenomenological temporality, rather than attempting to separate them. She clearly relies on both Aristotelian time, the present as central hinge for past and future, even if in a torqued version, and the present as a subjective and malleable duration. The book is sophisticated, and Ross's engagements with the field of temporality, with temporal philosophy, and with art are learned and well-articulated. But, and this is important for me, she specifically refutes Benjamin (whose work she articulates with admiration) as a philosopher of what she calls rupture, as well as other philosophers of rupture—a category into which of course Badiou also falls. It may be worth turning to Osborne, who specifies the difference between Benjamin's now-time (rupture) and both Aristotelian time and the durational present of the phenomenologists. He writes:

> Now time is neither the time of the blankly Aristotelian or cosmological instant, taken by historicism as the ontological ground of its chronologies and interpreted by Heidegger as the basis of the "ordinary conception of time"; nor, even less, is it the time of the extended, durational, phenomenological present described by Husser . . . . Rather, it aspires to condense into the punctual, one-dimensional space of the former the presence of history as a whole.[76]

But Ross is convinced that "suspension" of movement toward the future and new kinds of work with the present (rather than rupture) are the means by which we will best meet our obligations to the past (and its radical moments). As example, she cites Vivan Sobchak's argument that teachers cannot, in the classroom, make students care about historical pasts because they have not "lived" them. Therefore, teaching the past "must begin in the present."[77]

Ross herself acknowledges the perilous closeness of her "presentifying" to "presentism" as "a symptom of our times." She borrows the concept of "regimes of historicity" from Francois Hartog, who, in Ross's account, says that the fall of the Berlin Wall ends the narrative of the "futurism of modernity" since it is the end of a grand narrative. The result is that the present becomes the "privileged temporal category through which the past and the future are being . . . absorbed."[78] She says that for Hartog, "this present is a devouring present in relation to which the engendering of historical time seems suspended."[79] For Ross, however, "the diagnosis of presentism fails to account for the more creative ways in which the present is activated as an organizing principle of the past and the future."[80] And yet, her example of Sobchak's teaching seems to me a very definition of presentism: a truly conservative temporal force hidden in a conception of history that becomes, if anything, ever more distant from us as a thing that occurs without us and has little to do with us, a thing that we would not, and could not, dream of making our own.

I have left theatre behind for some time now. Let me return to a brief example, which happens to be a classroom, and about teaching, and is about a history that no one in the audience had "lived." It was true of at least one person with whom I spoke that this history had little relevance to her, and she wished the director had done more to relate it to the contemporary world (the present) so she could have found a hook. Guillermo Calderón's *Escuela*, which I saw performed in New York City at the Public Theatre's Under the Radar Festival in January 2016, is about a group of revolutionaries in Chile in 1987. The play is set in a series of classrooms in which masked young people are tutored in techniques for the revolutionary warfare underway to oust the dictatorship. It is a revolution school. Each classroom is devoted to a subject: firing guns or the economic mechanics of capitalism. The students are masked and do not use their names. What was notable and admirable to me was precisely that the play did not seek to establish any correspondence between this situation and the life of a New York City audience in 2016. It was decisively non-presentist, or presentifying, for that matter. It unapologetically showed, in the very simple framework of the classroom, a revolutionary idea being practised in full conviction. It addresses us as if we were in its own moment. This is exactly what Charles Isherwood objects to in his *New York Times* review. He refers to the play as a "dull, didactic drama" and comments that while "students of the Chilean revolution yearning to know what it was like to be inside the movement may find 'Escuela' engaging. I fear few others will."[81]

Without fully dismissing the play as Isherwood does, the person with whom I spoke nevertheless wanted the past dragged into coherence with the present. They wanted the past to be changed, in the sense of its terms made legible within an existing vocabulary of a present that seems to make itself the measure of the "relevance" of all that has come before. Our present is to be the measure of the way in which something matters. The artist, according to my interlocutor, should have created correspondences so that this revolutionary situation could become familiarized for us and we could thereby determine its value for us. Like some stage adaptations, this is a strange mimetic impulse to rob the past of its originality and make it imitate the present. It wants to enter the creativity (revolutionary energies) of the past into a string of likenesses, a series of events, all of which become recognizably similar. In the dialectical image or in the new present, by contrast (these are not the same, but I combine them here for rhetorical convenience), the present only exists to the extent that it is participating in the affirmation of the revolutionary energies, the energies of the "anonymous," in the past. It sees in its present, in a constellation, a past which does not evaporate into the conditions of the present, but exists in a fiery tension, a sense of ignition. We, in the present in this theatre, can feel ourselves to be in the present of that revolution, which is not like our times, but can live as our time, our present. Calderon does history in the theatre in a way that might suggest what Badiou calls a "rebirth of History, if by 'rebirth' is understood the emergence of a capacity, at once destructive and creative, whose aim is to make a genuine exit from the established order."[82]

2. Phenomenology and Theatre

Phenomenology is a branch of philosophy which has tended to attract more theatre and performance scholars than any other, including the foundational work of Bert States. It is an active area of investigation, prompting, for instance, the 2015 edited volume *Performance and Phenomenology: Traditions and Transformations*, and a large cohort in the Performance Philosophy research network. Because of its deep engagement with time, it is also, or therefore, at the core of one of the few monographs on temporality and theatre: Matthew D. Wagner's *Shakespeare, Theatre, and Time*. It is a book I admire in many ways. Wagner provides a set of analytic concepts drawn from phenomenology and then draws them through several of Shakespeare's plays. From the outset, he sets out to "revitalize our temporal sensibilities in theatre"[83] and claims that "the theatre has a habit of abolishing clocks."[84] Like me, Wagner wants to dislodge theatre from "chronos."

Phenomenology brings Aristotle's objective, cosmic time into relation with what Paul Ricoeur calls "the time of the soul,"[85] or what is also called subjective time. In Osborne's account of Ricoeur's work, the two times form an aporia, which phenomenological thought has attempted to resolve by working for a temporal frame that unites both. Aristotle thought of "lived time" as an effect of a person's awareness of the successive passage of objective time. Edmund Husserl, the philosopher who founded phenomenology, built on St. Augustine's three-part description of the lived experience of time to say instead that objective time comes from the experience of lived time. The experience of the present builds upon retention of what has come before, as hearing a sound now actually depends upon that sound being built across a passage of time, in what can then be understood as a durational experience, an extended now. In our experience of subjective time, a thick, durational now (including both retention and pretension—an anticipation of the future which also shapes the present), the succession of instants becomes secondary to the extended present that is the source of the retention that gathers a series from the past. Subjective time, very simply, is the feeling of time in which the sense of chronological time passing diminishes, in which the present seems to extend and thicken, in which our consciousness feels itself in a gathering of time that includes, but is also independent from, the sequence of successive instants turning continuously into the next.

Wagner, following Husserl and Heidegger, proposes theatre's unique revelation of time as this temporal thickness. For Wagner, theatre offers the displacement of our sense of chronological time by transposing beginning, middle, and end into a "temporal density" that we cannot feel in ordinary life, a "saturation of the present."[86] At the same time, this temporal density, this "inner time," collides, in the theatre, with objective time, making for "temporal discordance."[87] For him, it is precisely because theatre does not have to work, as philosophy does, at joining the two kinds of time that it is temporally creative. Wagner cites Rosalind in *As You Like It* as an example here. He writes that Rosalind's sense of a lover's time is more true than "clock time," and in fact does a better job of tracking time than the clock does. Wagner reads Husserl as saying that "to be human . . . is to be 'in time,' and to be in time is to be in an *immediate experience* of time, rather than a time mitigated by external measurement."[88] In his analysis, Wagner says that even if phenomenological time is "the more real time . . . it is objective time—and all its smoothing, straightening processes—that most of us live by. We are, by and large, simply better at living temporally consistent, rather than temporally discordant, lives."[89] But we like theatre, he says, because it does not smooth over time.

As I have said, for me there is much to admire in this book and I will not be able to give it its due here. But I have presented this core of Wagner's ideas in order to show that Wagner, and phenomenology in general, remains cramped within two ideas of time, one of which is personal and subjective, the other objective. It is as if there are only these two kinds of time. Between these two times, it is hard to see a place for the initiation of history. Wagner writes that cosmic or objective time is the one in which we locate shared social experience since it surpasses subjective temporal experience. At the same time, he seems to see objective time truly as "cosmic," outside of human history. And so it is strange to me, since it is cosmic, that he also designates it as the only medium for how collective and social organization (and therefore the historical) occurs. Wagner is clearly not here able to resolve this aporia between the two kinds of times so that social, historical existence can be accounted for. Osborne points out that Heidegger attempted to move past Husserl, while retaining his phenomenological core, in order to understand historical existence. But for Wagner and his view of theatre, temporal discordance wrought from the interplay of a historically untouched clock time and an inner "real" subjective and ahistorical time seems to be its own value, without any political or historical valence.

Bruno Latour offers a biting criticism of the adherence to only two kinds of time in his lecture/essay "Trains of Thought: The Fifth Dimension of Time and Its Fabrication." I use this essay at some length in Chapter 3, and so I will confine these remarks to his opening salvos. Taking a break from a conference of psychologists and phenomenologists in Neuchâtel, Switzerland—famous for its watchmaking—he strolls on along the lake watching a windsurfer. His irritation with the conference-goers and their adamant cleavage to "lived time" as opposed to objective time surfaces. He muses on the multiple complications of trying to claim, as he imagines his colleagues doing, that the windsurfer's experience (lived) would be ruined if his speed were calculated. They do not see that if his speed were measured, all the apparatus of measurement would also be part of the "lived world." Of lived time he says, "'lived,' one of these empty words that have no opposite, and are given a semblance of profundity because they appear to attack the cold and timeless and spaceless apparatus of dead reason."[90] He suggests to his colleagues who locate the richness of experience in the mind, in subjective perception, that instead they should look "to the world itself" in all its multiple entities and in all its processes, since "'objective' and 'subjective' time are like taxes exacted on what peoples the world, they are not all that these multitudes do and see and mean and want."[91] In place of these two terms, Latour wishes to "renew the philosophy of time," in which space and time are thought of as deriving from relations between entities. The difference he wants to make is to say that we should "generate as many spac*es* and tim*es* as there are types of relations." Further, "if other entities are necessary for our existence . . . then times and spaces will proliferate."[92] Such a critique is vitally important for my own project, as its premise undoes the central configuration of (two) time(s) that hound any attempts to think about temporality differently.

There is another aspect of Wagner's book that significantly contrasts to my work. This is the conviction that theatre is descriptive of, and experientially related and inexorably tied to, death. Here he works from his understanding of Heidegger and hypothesizes that, as with a life, the life of a performance takes on its unique identity by virtue of an ending, a death, that belongs particularly to it. Wagner writes that, while we do not necessarily know how a performance will end, we do know "that there *is* an end, and it has direct bearing on the 'now' of the theatrical present."[93] For him, theatre moves backward from death toward a "collective beginning" that happens as the performance begins, that relays us forward again toward death, or the end, through a temporally complex and rich present, or now, and back again. In Wagner's theory here, both birth and death move from something that remains primarily abstract and fictional in our actual lives to something "immediate [and] experiential"[94] in the theatre. For Wagner, one of the great values of theatre is that it makes death more real to us, more palpable, more certain, more the inevitable thing by means of which our presents unfold as they do. Here Wagner is locating himself squarely within those standard dramaturgical, structural, philosophical, and aesthetic emphasis on theatre's closures, its means of bringing its content and structure to a necessary end.

Linked to this disciplinary emphasis on death and ending, and also related to some extent to the influence of phenomenology, is a highly influential strand in theatre scholarship on time, in which scholars dwell on death, ghosts, hauntings, and memory as central to our temporal imaginings of and experience in theatre. Herbert Blau wrote, famously, that "of all the performing arts, the theatre stinks most of mortality."[95] This is a complex, deeply thought field to which some of our most admired scholars have contributed. The rich strands of surrogation, re-performance, and memory that thread through this field have been some of the most inspiring thought of the past decades. These include works from scholars such as Joseph Roach, Marvin Carlson, and Alice Rayner (with all the influence of Derrida's specter inscripted therein), not to mention Rebecca Schneider's *Performing Remains: Art and War in Times of Theatrical Reenactment*, which nudges memory, loss, history, and re-enactment toward new temporal thought. The interest continues, in part belonging to the Spectrality Studies initiated by Derrida's *Specters of Marx*, in volumes such as the 2014 *Theatre and Ghosts: Materiality, Performance and Modernity*. The editors of this volume, Mary Luckhurst and Emilie Morin, cite Jeffrey Weinstock on the emergence of the ghost as the "privileged poststructuralist academic trope" since "[in] either living nor dead, present nor absent, the ghost functions as the paradigmatic, deconstructive gesture . . . [the] trace of an absence" that can "suggest alternative, suppressed narratives."[96] Positioned within the fabric of postmodernism, the ghost's political and affective function is within what Badiou calls "the linguistic turn."[97] The political, the politically affective, has thus been seen to lie in the promotion of repetition, re-doing, cycles of replay, recycling, re-performance, ghosting, haunting, as if, far from initiations, there is nothing new under the sun and leftist desire has to content itself with small gaps, lags, little off-beats, and whiffs of spectral presences to reassure itself somehow that all is not completely lost. Notwithstanding the importance of this theatrical thought and the overwhelming evidence of theatre's long-standing imbrication with the revenant (which, as Carlson and others point out, is trans-historical and across cultures) and with death in general, I would suggest that a supplement (if not a replacement) to it is overdue. I work to supplement it, and specifically phenomenological work like Wagner's, with the life-affirmative, penultimate times of this book, times of initiation. My work in this book, especially in Chapter 3, is to free theatre (and temporality) of death-bindedness.

Another prevalent strand of temporal thinking in artistic practice and scholarship is durational art, which is of course also linked to phenomenology. In his book *Out of Now: The Lifeworks of Tehching Hsieh*, Heathfield formulates a theory of durational art and aesthetics in part from Henri Bergson. He writes that durational art work has been distrusted as anathema to the modernist investment in art as an event or rupture (which I have articulated as modernity's "new") or, as in Peggy Phelan's construction, an ephemeral "singular temporality."[98] As Heathfield writes, "but the model of thinking the temporality of performance as a shattering and recursive force is itself historically specific—part of the intellectual legacy of Modernity—and the predominant understanding of time within the terms of eventhood has tended to obscure other interpretations of the movements and forces of time, particularly the phenomena of duration."[99]

Heathfield points to performance artists who have for many years directed attention to time in the sense of the durational labor of the artists, of which endurance art is an important subset, and the risks or obligations of the spectator's own expenditure of extended time in watching the artist. He writes that durational work like Hsieh's "deals in the confusion of temporal distinctions—between past, present, and future—drawing the spectator into the thick braids of paradoxical times."[100] He writes, further, that "duration nearly always involves the collapse of objective measure. Whether it is short or long in 'clocktime,' its passage will be marked by a sense of the warping of time, an opening of regularity to other phenomena or inchoate orders."[101] What he pulls from Bergson is Bergson's sense of the "radical heterogeneity" of lived experience in the sense that in the duration of the present into which pasts are flowing we are never feeling the same thing, and attempts to spatialize or think that feeling of heterogeneity make it into a stilled object, a representation. What Heathfield looks to in durational performance is an "aesthetic perception" of the "sensate dynamics of temporality," both "as they are manifested in human presence [and in] the radical heterogeneity of durations."[102] Unlike Wagner, however, Heathfield draws this notion of heterogeneous duration into contact with a historically specific form of time, the accelerated time of capitalism. In fact, he locates the historical emergence of durational art in the 1970s as phenomenologically related to "capitalized temporality—regulation and acceleration."[103] In other words, the subjectivity of phenomenological duration seems to include subjectification within capitalism. In a blog promoting *Out of Now*, Tim Etchells of Forced Entertainment writes to Hsieh:

> If capitalism's structures are repeated in your work they are mirrored without the key element of productivity. It is different in each of the projects, of course, but in the "Time Clock" performance especially, the

> violence is felt instantly in the lack of any product at all (or any actual "labor"). There is only the submission or control of the body, time and space, the endless regulation and tracking of (human) resources.[104]

For Heathfield, "aesthetic duration" can "de-naturalize"[105] capitalist time regimes.

3. Influences and Innovations in Temporality (and Performance)

My research in temporality has been enriched by thinkers whose work is not necessarily included in the book, of whom I mention only a few here and only in passing. Furio Jesi in his book *Spartakus*, in his treatment of the Rosa Luxemborg and the Spartakus "revolt," gave me the means to compare and contrast the differences between revolt and revolution as forms of interruption of time also proposed by Badiou. For Jesi, for instance, revolt means "an insurrectional movement" that does not correspond to revolution. It is "a different experience of time" which does not depend on the long-range plans and consequences that help to constitute what he calls "historical time" and in which revolution takes place. Instead, in revolt, "everything has a value, independently of its consequences."[106] This rupture forms an instant collectivity outside of historical time, which forms a "shelter from historical time."[107]

Gary Wilder, whose work I draw from at length in Chapter 2, speaks of "anticipatory politics"[108] and provides keen readings of what he calls "relations to historical temporality," particularly with regard to futurity.[109] In Marx he finds someone who saw in the present "a future that is already in formation but is still difficult to imagine or discern,"[110] and in Ernst Bloch "a latent power congealed in emancipatory projects that are either 'not-yet conscious' or 'not yet realized.'"[111] He speaks of the "prophetic clairvoyance" in Aimé Césaire's poetry, in lines like "my ear to the ground, I heard tomorrow pass," which to me resonate with another guiding force in my thought on temporality: the poetry of the Palestinian poet Mahmoud Darwish and lines like "Maybe now is much more distant. Maybe 'yesterday' is nearer / And 'tomorrow' already in the past / But I grasp the hand of 'now' that History may pass near me."[112] A poem by Langston Hughes called "Scottsboro" has been a constant accompaniment for me and for my conviction about the political vitality of the concept of the new present. Kristin Ross's wonderful book on the Paris Commune provides me with another affirmation of this idea when she writes about the Commune's replacement of "*mesdames* and *messieurs*—the bourgeoisie *les honnêtes gens*"—with the self-description *citoyen*. While *citoyen* "may well be old and originate in another moment of the political past," its iteration in this instance creates the now of a shared political subjectivization, "the uncomfortable class struggle in the present."[113]

Of course, a great deal of rich, productive, and sustaining work on temporality has been done in other disciplinary fields such as Media Studies, Art and Visual Studies, and Queer Studies, far more than in Theatre and Performance Studies. While these are for the most part outside of my particular conceptual terrain here, they are not outside my fields of interest and engagement. For instance, Ross's survey of methods and strategies by artists working in temporality is most helpful, even as she, as I have already said, holds temporal malleability to the present, to presentism, and rejects temporal theories of rupture. She describes, for instance, a set of methods used by artists that "significantly alter the future-driven modern deployment of historical time"[114] without rupture. These include "freeze, reorient, reshuffle, flatten out, lateralize, spatialize, excessively hold."[115]

I have been stimulated by critical analyses of neoliberal time, some of which is included in Chapter 4, and which includes the challenging work of accelerationist theory. As I discuss briefly in Chapter 3, Lauren Berlant's brilliant and guiding text *Cruel Optimism*, for instance, describes the temporalities of affective genres of day-to-dayness and survival in neoliberalism, in a text that is grueling in its affective despair. These circulate around a more and more impossible to sustain optimism as a defeated genre of the liberal life, with optimism as a certain relation to the future, a certain living in hope of escaping the indignities, precarity, and immiserations of the present. Her work only increases my conviction about the urgency of allowing performance to help us conceive extraordinary times, to be sensitive to political, material, fiery matters that do in fact keep exploding into appearance without relation to futurity.

Theatre and Performance studies have begun to undertake temporal studies in earnest as I write in mid-2017. We have seen working groups, conference panels, journal articles, edited volumes, monographs, and monograph proposals emerge, a new plethora of innovative thought that disentangles temporality from processional history. Much of it, like mine, delinks from the future, working against it in subtle ways torn from processional history, but without Ross's presentist, or presentifying, overtones. For instance, in an excellent essay Guilia Palladini writes that following the 2008 crisis in Germany, the country is haunted by the return of the Weimar Republic and its crisis, which is deployed by conservatives as a warning. What interests Palladini is that this return of the Weimar may bring that twelve years suffused with revolution "yet to come" with it, its "projective temporality."[116] It is possible, she says, to "reclaim" the Weimar for "a radically different politics of use—one in which the anticipatory logic of revolution inhabits the present territory of potential."[117] Key to her argument is a sense of "rehearsal": that certain forms of solidarity can be tested "in the present, even if they have no immediate or precise political outcome."[118] The future is projective only, the work of the present, which is infused with the revolutionary temporalities of the past. She figures the unemployed worker of the Weimar with the precarious worker of neoliberalism, each engaged in "passionate activities that act as a prelude to revolution" in a future that is not at the end of a linear process but in a to-come, just as it was in the Weimar period.

The excellent Winter 2012 edition of *TDR: The Drama Review* on "Precarity and Performance," edited by Nicholas Ridout and Rebecca Schneider, contains essays that in many ways circulate inside this book even if not directly cited. In particular, I have been inspired by another essay by Palladini and her proposals about idleness and the temporality of foreplay, Rebecca Schneider and the idea of the temporal lag, and Tavia Nyong'o and his meditation on kairos (with kairos at the core of Chapter 4) as "intensive, revolutionary, occupied time."[119] A special issue of *Performance Research: A Journal of the Performing Arts* emerged from the 2012 Performance Studies International (PSi) conference at Stanford University called "Now Then: Performance and Temporality." This conference was most influential to me, and in a sense launched this book since it was there that I saw Cassils perform *Tiresias*, a performance I wrote about in an essay in *TDR* and that is at the heart of Chapter 4. The special issue, introduced by Branislav Jakovljevic, is a wonderful collection of essays, two in particular of which remained with me. One is Peggy Phelan's lovely essay in which, followed by a study of the human microphone used during the Occupy days, she cites Agamben on changing time as the truly revolutionary task and says, "Live art is an art of time, and therefore if we are looking for revolutionary thought and action, performances that alter time in one way or another may have potent political power."[120] The other is Natalie S. Loveless's essay on ice time, which, although located firmly in a phenomenological treatment of durational performance and drawing in large part from Heathfield, asks good questions, ones oriented toward "multiple ecologies," and in particular the following, which stays with me on a daily basis: "How many things in your immediate environment alone could you temporally coordinate yourself with? Why live by one time alone?"[121]

Two books are of particular importance as part of an emerging field of temporal studies in theatre and performance, one of which I have already mentioned: Rebecca Schneider's *Performing Remains: Art and War in Times of Theatrical Reenactment*, and Nicholas Ridout's *Passionate Amateurs: Theatre, Communism and Love*. While this latter is not specifically about temporality, Ridout, guided by his revival of romantic anti-capitalism, locates a communism within capitalism that is especially traceable in the relations of labor and time found in the theatre, in the difference between professional and amateur labor. While my own work diverges from Schneider's dazzling conceptual wordplay with the (re), her study of re-enactment is a complex innovation in many of the assumptions about theatre and performance's relation to temporality that finds, over and over, that theatre's temporality is "volatile, easily swerved."[122] She says that "the queering of time troubles our heritage of Enlightenment investments in straightforward linearity as the *only* way to mark time, reminds us of a durational 'now' for political action, and points to a politic in veering, revolving, turning around and reappearing."[123]

This overview of influences and temporal innovation cannot possibly do justice to the complexity of the work that I have surveyed so quickly. But I hope that it will provide my reader with a sense of the overall terrain, as well as further orientation to my own position and arguments.

Methodology/Positionality

This book is performance driven. That is, unlike my previous books, theory and thought in this one weave in and through performances (and theatre texts), in theatres and on the page. I attempt to describe these performances intimately and densely and in the ways they have helped me locate how radical temporal theory may manifest in performance. As such, the performance descriptions are a mix of an attempt at communicating performance elements "objectively," and my seeing, my thought, my investment in temporality becoming entangled with them. Performances are never mere objects of study for me. Always, if I like them, they are too vibrant for that. As for most theatre and performance scholars, I imagine, they affect me, excite me, cause chains of association, and instigate sharp and sometimes exhilarating intellectual arrivals. It is in this interwork with performance that the work in this book happens. I have endeavored to weigh the book in favor of this deep engagement with performance and to keep to a minimum the presentation of philosophical thought, or the full complexity of an idea. I have endeavored to provide, in as simple a way as possible, only what is needed. Nevertheless, the reader will still find, and perhaps already has found, passages that require some work parsing through philosophy to get back to the matter at hand. I hope to guide readers well.

My three central models for thinking about time in this book are European (and white and male), and may perhaps be brought to task as such. These choices are highly selective, and my scope is relatively small. My aim in this book is not to parse my choices through a testing of them against the entire oeuvre of Western philosophic thinking on time, against Kant, Heidegger, Nietzsche, Bergson, Husserl, Ricoeur, and so on. My intention is to be performance-bound, rather than philosophically exhaustive, and draw from a limited set of temporal thought that is for me the most provoking, radical, and rich in providing guides for political practice of as yet unknown or uncertain forms as they occur in theatre and performance.

It is also true that the European binding of this work is highly exclusionary of diverse temporal thinking produced and being produced everywhere else in addition to Europe, including in indigenous contexts and activism, in Afrofuturism, in the speculative theory of the kind of thought that is radically pushing on all received notions of environmentalism and save the planet lingo, as in Donna Haraway. (The last three will be briefly introduced at the end of Chapter 4.) And it is also my intention to at least begin the work of citing the generation of some kinds of temporal thought in, for instance, black modernist, internationalist, and communist thought as they influenced European thought. Despite the apparent European selectivity of my choices, my own thinking is attentive in these studies of temporality to, as I have said in my definitions of initiation and history, possibilities to inaugurate universal (belonging to everyone and excluding no one) conditions of justice, recognition, and equal distribution in particular local and/or globally related situations. Ultimately, I do not think that the European binding of my work here is as exclusive to the white, male European as some might feel it is. Or, at least, I am personally invested in ongoing research into the ways in which the forms of temporality that are central here are also echoed in, fragments of, in collaboration with, other possibilities for radical temporal thought that do not have their origins or practices in Europe.

I am aware that Badiou in particular can be an irritant and is often criticized, especially but not only in theatre and performance circles. I of course find blind spots and shortcomings in his work, but I do not take the time or feel the need to defend him here. He is well able to take care of that himself. The work of Negri has been more popular, especially as it has been dispersed through his work with Michael Hardt in the volumes *Empire* and *Multitude* and in its intersections with the always popular work of Deleuze and Guattari. It has been perceived to have more of a relation to a real-politic than that of Badiou or Agamben. Agamben—his work mysterious and difficult, profoundly and densely etymological, and very often sourced in Judeo-Christian texts—seems to me to be the most removed from "the ground," from the conflagrations of destruction, immiseration, race, and speculative capital. Badiou, the most temperamental of the three, is passionately attenuated to these conditions, and, without necessarily extending his thought to real-world examples, clearly founds his urgent need to philosophize in them.

It may be objectionable to some that the words "we" and "us" appear throughout a book based in white, male European philosophy. Again, my concern is to see in that philosophy the possibilities for a distribution of justice. When I use the word "we," I am inviting my readers to participate in an imaginary collective that will have a standpoint related to justice. I of course acknowledge the absolute and definitive differences of positionality, race, class, experience, and so on in both my readers and in the people I may be referring to in the work. Nevertheless, I choose to make a gathering gesture, a gesture toward speculative shared work and real positioning within current temporal regimes. Those regimes touch all of us.

It should be said that for me in this work, performances and my thinking are not directed toward any political efficacy. I do not predict or suggest any political outcome. My intention is speculative, directed to a speculative politics generated in some theatre and performance through its material means of embodying alternative times. It is to suggest times in the theatre that, if we are attentive to them, can ignite those times with the imagination of other times than those by which we are too tightly bound.

I have called this introduction Chapter 1 because, in addition to a survey of what I intend to do, it contains a considerable amount of substance that is part of the book itself, or the book needs the outlay of this considerable substance before the work on specific performances can begin. Chapters 2 and 3 are followed by a shorter Chapter 4, which will also provide a sort of coda to the work. In some cases, throughout the book, the quotations from performances are from notes I took while watching them, rather than from published texts. These are *Before Your Very Eyes* and *Real Magic*. Therefore, if I do not have endnotes and citations for quoted phrases when discussing these performances, please assume that they are from notes. I have also consolidated some endnotes, so if there is not an endnote number at the end of a quote, assume it is from the source and page number cited in the next endnote.

The Chapters

Chapter 2: There Are No More Slaves—The New Present and a Temporal Philosophy of Revolt (Plays on the Haitian Revolution)

In an article on the Nuit Debout movement in France, Gabriel Rockhill writes the following:

> When there is attention paid to radical insurgencies, the mass media, professional politicians and well-paid pundits revel in stories with clear beginnings and ends, thereby securing closure in terms of a simple narrative logic that commonly juxtaposes dawning aspirations to dusk-like disappointments. In the beginning, we are frequently told, there was light: individuals in a specific location like Paris suddenly "awakened" one day in order to come together in a collective act of protest. After battling for a specific goal and attracting the supposed daylight provided

> by corporate media coverage, they then sink back into the abyss from which they came, perhaps leaving things worse than before. Spectators are subtly encouraged to wonder why there was even a protest in the first place, the implicit message being that it is best never to try and change things at all (since they are ultimately unchangeable). . . . The space and time of insurrection are thereby divided and conquered. . . . They are then inscribed within a temporal framework of original aspirations and final consequences. Everything becomes a question of means and ends: what is the goal and was it attained? The only possible success is thus defined in terms of "productive" results within a delimited space and time, as if the revolutionary transformation of society *in toto* could be reduced to the same logic as capitalist profit margins.[124]

In this chapter, Badiou's concept of the new present is developed as a strategy for engaging in the absolutely necessary combat with both liberal and leftist tendencies to "divide and conquer" the "space and time of insurrection." The plays under consideration here, in quite different but linked ways, interrogate and present the Haitian revolution. Conditioned by quite different historical situations, with the first written nearly thirty years prior to the second, they are *Toussaint Louverture: The Story of the Only Successful Slave Revolt in History*, by C. L. R. James (1935), and *The Tragedy of King Christophe: A Play*, by Aimé Césaire (1963, rev. ed. 1970). Encouraged perhaps by Césaire's title, scholars have framed the plays so that the revolution the "tragedies" enact is enclosed in a heuristic net that must, given the genre, designate the revolution failed. Its leaders, tragic heroes, suffer the fate of all tragic heroes.

The new present, by contrast, is a tool with which to disengage from this habitually practised tendency to see in past revolutions, movements, insurrections, in Occupy Wall Street, only their predictable failure. The new present is a temporal tool for changing the heuristic model that inevitably finds the impact of uprisings inconsequential. The new present abolishes the tragic implacability of the progression toward destruction and ruin. Finishing her chapter on C. L. R. James and W. E. B. Du Bois in *The Intimacies of Four Continents*, Lisa Lowe proposes what she calls a "past conditional temporality [which] suggests that there were other conditions of possibility that were vanquished by liberal political reason and its promises of freedom, and it suggests means to open those conditions to pursue what might have been."[125] I take these plays precisely as a way to experiment in assuming for theatre the task of opening vanquished conditions in their new presents.

The Haitian revolution itself is emerging into a new centrality after being obscured in both liberal history and some historical analysis within black international and black radical thought. Attention to this revolution as pivotal reveals an internationalist force field of the foment of ideas of equality, revolts, and insurrections by enslaved and other oppressed peoples, circuits of inspiration and information traveling intercontinentally as what Peter Linebaugh and Marcus Rediker call, in the subtitle of their book, *The Many-Headed Hydra*. The hydra is a body made up of so many points of insurrection, planning, gathering, and organizing that cutting down a part of it cannot stop its strategically and brilliantly evasive entirety. Embers of insurrection traveled during the centuries of circum-Atlantic plantation slavery via revolutionary bodies, the sea, newspapers, revolutionary organizations, ports, and preachers, toward sites of their re-ignition. History becomes the flares on the boats in the harbor, presaging the uprising, while behind them, in the oceanic blackness of the Atlantic, there are other flares, receding like stars whose light still arrives in our midst.

Attention to the Haitian revolution, from C. L. R. James's early analysis in *The Black Jacobins* to now, also contributes to the critical assertion that slavery was foundational to the development of capitalism. Further, it also revokes from France and America the legacy and primacy of revolutionary thought and action during the years in which the American, Haitian, and French revolutions took place. In some senses, the Haitian superseded the other two, since only the Haitian revolution resulted in slave emancipation and a true foundational change to the notion of freedom. In other words, what gets offered to the world when we start admitting the Haitian revolution into general historical consciousness is nothing less than a remake of the world, a newly vital sense of who those people have been who initiate its trajectories toward equality. As such, to take up two plays about that revolution, plays that study its internal contradictions, promise, brilliance, and nearly crippling choices, is to recuperate for theatre and theatre scholarship the world that the Haitian revolution opened and continues to open. A world which, in that new present, "cancels the slave ship,"[126] in the words of King Christophe in Césaire's play. The slave ship needs canceling again and again. I am interested in how these plays, at whose core is the "Black Spartacus," are in themselves embers.

Chapter 3: Changing Time in the Time Before the End

The Gob Squad's piece *Before Your Very Eyes* opens with a group of children and a question: "'You know why we're here, don't you?' We're here to live and die." The question refers, mischievously, to what will happen in the show, but also functions as a double-entendre in reference to our own lives, those of us in the audience. It is also about the theatre itself, its presumed life-cycle as it moves in each performance from the beginning to the end, sometimes leaving dead bodies in its wake, generically wrapped up in death.

This chapter forms the book's central disputation with death. That is to say, not with the fact of death itself, but with the narrative essence that is often inseparable from it. This is the sense in which life is a course toward death, its inevitable theft of life ruling over an unalterable tick tock countdown, as in Richard's "For now hath Time made me his numb'ring clock."[127] Our hearts, our ticking time bombs. Our breaths, numbering our days, our bodies metaphorically and physically wrapped into death's oncoming-ness. A sense of biological, social, and artistic finitude weighs on our sense of our own time, and what we can do within the ticking out of our days.

I have become intimately familiar with death in the course of preparing to write this book. I held or stayed close to my husband's dead body in the emergency room for 8 hours after he died without warning, at home one morning, when his heart skipped a beat, or beat too slowly, or beat too fast, went out, somehow, of its orderly, stately, procedural course. Eight months earlier, my father had died, at the end of a long and rich life. My closeness to him throughout my life was immeasurable. Grief has become a constant in my life. I am baffled at death. I am baffled at the inability of the human mind to really believe it, especially in the case of a sudden death. It is impossible for it to become true, even as every day one lives it as true. My heart stretches, without its measuring beats, across the world to everyone who is enduring, day after day, the death of people they love, the endangered everywhere, people targeted for death and those who mourn them.

It is even possible, as Judith Butler writes, that "loss has made a tenuous 'we' of us all."[128] Her powerful work on grief and vulnerability, in which by virtue of these we are "given over to the other"[129] in a way that could presage a new politics, has been a powerful incitement, at least in scholarly and activist communities, to think of politics as/through mourning, a relation to death, a response in the face of its sheer magnitude across the world among the vulnerable, those who are not even considered mournable.

And yet locating politics, or creating a hermeneutics of life, including in the theatre, from mourning loss, from the ineradicable fact of death, is countered in this chapter by a powerful reorientation to death, to endings, in Agamben's idea of the penultimate. The penultimate is the time before the end, which Agamben suggests is a time within chronological time in which it is possible to make time our own. Penultimate time is not determined by a course toward death, a linear procession, as is chronological time. Agamben differentiates between the two by calling the penultimate time "the time *that* we are," and linear procession as "the time *in which* we are."[130] The time in which we are is that helplessness of being borne along with time, of being almost a spectator of ourselves passing our lives in time's current. The time that we are, by contrast, is the sense of initiation. It is a taking hold of time so that it can be changed. It makes linear passage of time three-dimensional, expanding it into something in which we can make our own time. It is an affirmation of the possibility of a creativity of time that produces an existence not reducible to death, which may mourn at times but seeks more continuously ways to, as William Kentridge says, involve itself in the "creating that comes before [death]."

Kentridge's opera, *Refuse the Hour*, is one of the four performances that comprise this chapter's work. The others are *You Are Nowhere*, by Andrew Schneider, *Before Your Very Eyes*, by the Gob Squad, and *On the Concept of the Face: Regarding the Son of God*, by Romeo Castellucci.

You Are Nowhere, the first study, is an extraordinary piece in which Schneider, more literally and actually than any of the other studies in the book, changes time. Repeatedly framed as going back to the time before death, he uses a mad mass of precise technology, physics and thought experiments, falls, disruptions of the smooth flow of theatrical time, accomplished physical and vocal work, and pure shock to create exactly and thrillingly the time that we are, the time taking place in but extracting itself from *linear* time. It is a seizure of time, and the at-times inchoate and always difficult work of creating a different time.

Refuse the Hour is a joyful affirmation of the creativity of penultimate time, performed on a crowded stage full of invented time instruments (including musical instruments), by singers, musicians, and the dancer Dada, who is in this piece a kind of co-creator with Kentridge. Dada is featured in the photograph on the cover of this book. The piece is framed explicitly as "not resisting mortality in the hope of trying to escape it, but trying to escape the pressure it puts on us."[131] This is linked throughout the piece to the attempted imposition of European time on an increasingly colonized world, particularly in Africa, which is refused, performed onstage and in short films as colonial revolt. I pay particular attention to Kentridge's philosophy of the studio. During the performance, films are shown at the rear of the playing area in which Kentridge, often with Dada, is at work in his studio, work he describes as attempting to change time. I also pull Bruno Latour through the chapter for his re-do of the classic thought experiments of the passenger on the train, and the observer, and the twins, one of whom is left behind when the other travels speedily away. Kentridge uses both, and Einstein's passenger on the train is also reworked by Schneider.

Kentridge's penultimate time becomes full of blown up trains, multiple entities, being left at the edge of the black hole, revolt, defiance and joy, and, finally, Kentridge's signature filmed procession in silhouette across the back of the stage, toward the black hole, at the same time as singers and musicians in full color and vibrant life form a procession which moves downstage toward us, in what is an initiation of history claimed by veering from processional history into penultimate time.

Before Your Very Eyes turns us to children. In this piece, a group of children in each city in which it has been produced has worked with Gob Squad company members over the course of two years in order to document changes in their aspirations as they age. In performance, they appear in a clear box, lined with one-way mirrors, so that we can see in, but they cannot see out. We see, on film at either side of the stage and the box, films of each of their younger selves. There is a disembodied voice which directs the children, beginning with "You know why we're here: we're here to live and die." The explicit purpose of the piece, it seems, is to stage exactly that progression/procession toward death that this chapter refutes. The children, in hokey, dumb costumes, enact scenes from stages of life determined by the voice. Their experience is one of increasing disillusionment and disappointment, ennui. In the last scene, they act out old people and are told to stage their deaths. My argument in the chapter is that this is not a portrait of the inevitable decline from youthful dreams and sense of immortality to the certainty of finitude, but something else entirely. Here my thinking begins to be directed in part toward the work of Chapter 4 on the concept of the future. The children in *Before Your Very Eyes*, I argue, are withdrawing from their role as the future. The children are relieving themselves of their futural obligations, and as such their obligations to processional history, which projects children as ongoingness. And it also relieves them of the obverse, the sense that children too will ultimately succumb to a sequence of life disappointments ending only in death. Because here the children, after staging their death scene in their Halloween-style age makeup in exaggerated child versions of heart attacks and strokes, dance back from their deaths, literally dancing backward via the screens into living joy. I draw on Walter Benjamin's essay theorizing how children of the proletariat should be educated through years of theatre work, in a way that removes both communist and capitalist futures as the directive for this training and replaces the future with to-come, a time concept that will be central to the next chapter. I work to illustrate that the children in *Before Your Very Eyes* are literally using the theatre to create an operational, chronogenetic time that defies not only processional time but also the habitual assumption of theatre's inextricability from death and endings. They take hold of time.

On the Concept of the Face: Regarding the Son of God stages the progression of an old man's deterioration toward death by way of increasingly uncontrollable incontinence, at times literally flooding the stage with his diarrhea. His son, in this pristinely white house, attempts to clean up after him. Massive, spread monumentally over the back wall of the performance space, is a fifteenth-century portrait of Jesus with a particularly compassionate gaze. Eventually, the old man, finally alone, sits on a bed on which he has spread his own waste. A child appears, with a backpack, from which he withdraws a grenade that he hurls at the face. Gradually more children appear and likewise hurl grenades, producing a cacophony of sound, of destruction. They arrive in the old man's penultimate time, insurrectionists, stone throwers. They make a chronogenetic time, a time where time is being made, insubordinate to processional history. They are perhaps the clearest example in the book of the initiation of history. When they are finished, the old man does not die, but gets up and walks offstage. The face is destroyed. The play begins with the certainty of death, of the procession to death, but there is no death in the play, nor is there an expectation for the future figured in these children. There is just the destruction of two time images, theological and mortal, which try to guarantee submission to a time in which we are helpless.

Chapter 4: The Volatility of Time in the Hold: Kairos and the To-Come

This chapter picks up centrally on the idea of the to-come, or kairos, as a disorganization of and replacement for the future. Each of the pieces, in different ways, provides a plastic, topological, theatrical architecture of being without the future of processional history. I work here primarily through a loose adaptation of Antonio Negri's conception of the to-come, time on the brink of time, a hold, leaning out over what may be, but without the futural directionality of processional history. Throughout I work with an idea of a hold, or an immobility, an idea generated from a performance lecture given by Tim Etchells of Forced Entertainment, which I extend through Stefano Harney and Fred Moten's essay "Fantasy in the Hold." It is an idea of creative irresolution, creative being, energy, the uncapturable, subtracted from futurity and, in particular, modernity and hypercapitalism's different forms of futurity.

The future has in some ways become especially singled out in the late twentieth and early twenty-first centuries' formation of hypertrophied capitalism. That is, in the time during which the readers of this book live, processional history has to some extent become convoluted, turned back around on itself. The future becomes a speculation that lives in and distorts the present and becomes *the* anxious, nauseating thing that adjudicates it. None of the fundamental features of processional history or the chronological time that conditions it have actually changed; it still depends upon being immune to revolt, on the suppression of temporal initiation and the initiation of history, and it is still founded on injustice. But it is important in this book that hypertrophied capitalism's torqueing influence on classical capitalist time, especially of the future, is acknowledged and met with a portrait of temporal initiation that specifically releases the hold of the future as a category in any of its forms.

The performance examples here, with the exception of one, are not about hypertrophied capitalism and its forms of the future. Rather, I am situating their allure within the perspective of an audience person who is living in and through those forms. The future, in our time, manages us, creates us. It is not we who will make the future, but the future that has already happened to us in that we bet on it as if it had already taken place. Its tricky folds and twists are engorged with our small speculations, our anxiety. Of course, this has been happening since the beginning of speculative capitalism, which was also the onset of racialized capitalism, in which a new insurance industry speculated in the lives of slaves on the basis of the profit that might accrue from their (often probable) deaths at sea. But it has become even more systemic and accelerated. This is the situation that makes it urgent to imagine and initiate a temporality that holds at the brink of time and therefore, according to Negri, is the point of innovation and creation.

The chapter begins with *Real Magic* by Forced Entertainment. I read the piece counter-intuitively, since in many ways it is a portrait of the kind of hypercapitalist time described in the accelerationist theory that I both use and critique here. It is that time in which the future almost appears not to exist, so turned is it into risk and precarity in the present of our daily lives. It is also, however, an experiment with energy in the hold, an explosion of energy that, while not kairos, is a kind of containerized protest in which a different kind of political imagination surfaces.

This is followed by a look at Cassils's solo performance *Tiresias*, another image of the hold, which is what Agamben, in reference to a Bill Viola video, calls "kairological saturation."[132] Cassils clearly draws on the category of the future in using Tiresias, not only as someone who lived as both genders but also as the blind seer. In the piece, Cassils is pressed, nearly naked, into the negative side of a cast of a Greek male torso made of ice. They stand, still, holding, as the ice melts. I argue, starting from Cassils's own articulation of the piece as holding at the irresolution of male and female, that this is a moment of kairos, a not-moving forward into resolution, but a leaning out over the to-come of what might be. Just as Negri replaces the future with the to-come, he replaces the past with "the eternal," as Badiou replaces it with the new present. Negri does not mean by this a transcendent, a-historical category but an accumulation of vibrant creative kairos. Cassils holds us in the place of kairos, where divination is not related to speculative futures but is alive with kairos, something new, and uncapturable.

I move from Cassils's performance to a reading of the final forty-five minutes of Carl Theodore Dreyer's *The Passion of Joan of Arc*, another image of immobility, fire instead of ice. I argue that Joan also holds at the edge of time, in a kairos that stops the violent brutality of the church as it attempts to prolong its power into the future and erupts into fiery revolt against violence and those who have the power to administer it. I compare this with another piece by Cassils, *Inextinguishable Fire*, in which they are set on fire via a Hollywood stunt technique. They are in a hold for fourteen seconds, coming to an edge of time beyond which they will be harmed, suggesting perhaps the power of self-immolations as fiery initiations of kairological time, a way to decline the future and instead lean out over a to-come of some alternative existence, still indeterminate and unresolved, but in revolt against the ongoing violence of processional histories.

Finally, I turn to a dance theatre piece by the Lebanese choreographer Omar Rajeh, which I treat as an affirmative hold, a kairos of joy. In front of a ticking clock high up on the back wall of the theatre, Rajeh and three other dancers, accompanied by three Palestinian oud musicians, alternate between dancing and chopping the vegetables for Lebanese salad at stations along an extremely long and beautifully laid out, gleaming metal chef's table, at which an older Lebanese woman chops and cooks continuously. The fragrances of mint, parsley, onion, cucumber fill the air. It seems that they are cooking and dancing an extraordinary time into being, disjunctive from the chronological time of the ticking clock. What the clock says has no impact at all on what is happening in the mix of dancing and cooking. There is just the initiation of the dance, an improvised undertaking by each dancer, out of the collaborative merriment of the cooking, which seems to have no particular directionality, no beginning or end. And gradually, the clock starts moving backward and starting over, never quite coming to an end of ticking away the minutes of the performance. The making of the food is clearly future directed (when it is finished we all eat it), but we are somehow not in the time that is passing but in a kairological immobility, a fullness of active creation to which the clock itself eventually surrenders.

Finished with these performances and an exploration of kairos, the rest of the chapter is a summary, a small extension, an invitation, and deliberately ends with the challenge of the to-come, as the book's resistance to its own finality.

2

There Are No More Slaves: The New Present and a Temporal Philosophy of Revolt (Plays on the Haitian Revolution)

This chapter is concerned with two plays about the Haitian revolution: *Toussaint Louverture: The Story of the Only Successful Slave Revolt in History*, by Trinidadian C. L. R. James (1936), and *The Tragedy of King Christophe: A Play*, by Martiniquen Aimé Césaire (1963, rev. ed. 1970). I bring these texts, and the revolution itself, into contact with Alain Badiou's concept of the new present. The new present is a difficult tool to grasp and use, perhaps most especially because its radical disputes with historical and temporal normativity are so novel. This chapter seeks to establish it as at least a necessary component of a retooled and refurbished toolkit for thinking about both historical and contemporary revolutions, revolts, uprisings, riots, and emancipatory occupations, especially with regard to theatre and performance. In the course of this chapter, that toolkit will also be outfitted with work from Black International thought and the Black Radical Tradition that intersects with the new present in provocative ways. A second thread of the chapter will be to question the genre of tragedy and its temporal conditions as it has been applied to these plays, and to revolutionary failure in general. To frame the new present in relation to the plays, it is simultaneously necessary to rescue them from the finitude of tragedy.

The Haitian revolution (1791–1803) resulted in a historical initiation and novelty; for the first time in history, slaves freed themselves from slavery. They did it within the context of viral and vicious plantation slavery in the colonial Caribbean and the newly postcolonial United States, *and* as part of preceding and concurrent collaborative constellations of slave uprisings and revolts across the Caribbean and the United States, or what Nicholas Mirzoeff calls "the Atlantic revolutions."[1]

It was also the first time that former slaves were, as Eugene Genovese says, "struggling for access to state forms,"[2] a struggle that resulted in the establishing of an independent state under the leadership of former slaves. It was also, for C. L. R. James and Césaire both, the site at which the idea of European centrality was overturned, as Black Jacobins formulated and practised an idea of equality and liberty that pushed on and informed the French Jacobins in the concurrent French Revolution. As Nick Nesbitt writes, Jacobinism was the "the militant defense of the idea of undivided equality" and Black Jacobinism, "in turn, would describe the ways this defense of equality and popular sovereignty were displaced and transformed, tropicalized and radicalized, by a series of figures who together constitute a francophone, Black Jacobin radical anticolonial tradition."[3] For Mirzoeff, the two revolutions are absolutely intertwined, in that "urban radicals" (the *sans-culottes*) and "the enslaved" "revolutionize[d] the [French] revolution."[4] Each of these fought centrally for what amounted to the same issue: "sustenance and sustainability" for the revolutionaries of Haiti and "the right to existence" for the French revolutionaries. And both, Mirzoeff continues, "visualized themselves not only as acting within history, but as making History."[5] In acts of historical initiation, "the enslaved had renamed themselves as a people, classified themselves as rights-holders, and aestheticized their transformation in the person of the Hero [Toussaint] and the concept of History."[6]

The Problem with Tragedy

For me, then, the plays about this revolution can be understood as sites of temporal innovation and the initiation of history. That is, the texts can be studied and embodied with an eye to this innovation and initiation. It is my central contention that the Haitian revolution, its present, comes into existence, again and again, in different worlds and different conditions—the definition of the new present. As such, I am resistant to the coding of the plays as tragedies by scholars who seem to attach the tragic genre to the Haitian revolution and to revolution in general. For instance, anthropologist David Scott, who lived through the catastrophically failed Grenadian Revolution, describes tragic temporality as "stricken with immobility and pain and ruin,"[7] in dissonance with the promises of revolutionary narratives to unfold the revolution into the future. For him, tragedy is a genre that can help understand the lived experience of people who live on "in the ruins,"[8] in the disastrous endpoint of that once-imagined historical unfolding. I have no desire to sustain the idea of revolutionary progressivism, and in fact counter it throughout this book. But in borrowing the tragic genre to support an idea of the necessary finitude and death of revolutions, including the implication that existence is making do in the ruins, Scott mobilizes the genre for something that to me seems to be conservative modalities.

Jeremy Matthew Glick, on the other hand, in his sophisticated and convincing arguments on the use of tragedy by James, revisions the tragic, for one thing, as a formal means by which to study how the relationship between the leader and the people unfolds in a revolutionary situation like the Haitian revolution. In general, he suggests that genre categories such as tragedy need to be modified by new historical conditions. To the extent that James and Césaire did draw on classic tragic structures, and, in the case of Césaire included the word "tragedy" in the title of his play, the genre was modified by communist conviction.

As opposed to those critical viewpoints that would, using these theatrical treatments of the plays, sentence the revolution to a finite sequence at the hand of failed tragic heroes, Glick writes, "Haiti is the generative site *par excellence* for creative work by African diasporic artist-intellectuals . . . . Revisiting Haiti acts as a solvent against political ossification."[9] He implies here a vibrant and vital presence in new worlds of the capacity for political invention that appeared in revolutionary Haiti. I use the plural form of worlds throughout to draw from Badiou's conception of the existence of many worlds at the same time. When I say new worlds here, I mean the international or trans-geographical and radical political scenes to which both Césaire and James were committed, and within which, as Gary Wilder says about Césaire, they "resisted the idea that they should approach modern philosophy as foreigners."[10] Each found in Toussaint a form of political thought that worked against the resolution of colonialism in an independent state, and envisioned a federalist system which ensured cooperation, exchange, mutual growth along with autonomy, the guarantee of the end of slavery, and equal citizenship. Toussaint, the Black Jacobin coordinating the slaves become Black Jacobins (as James called them), sought, in the French Jacobins, the terms of a mutually constitutive formation of equality and liberty. And, far from succumbing to revolutionary failure and finitude, both Wilder and John Patrick Walsh, among others, pay specific attention to the complexity of temporal thought in Césaire's work. Here, for instance, as Walsh writes, Césaire thought that "the past contains within it a potential that must be reactivated in the future,"[11] an idea very close to that of the new present.

However, after a chapter in which he discusses Césaire's influential essay on Toussaint, through which he is able to recount Césaire's temporal innovation, Walsh turns to the theatre, to *Christophe*. It is at this intersection with theatre that, for him, the tragic becomes the central hermeneutic for thinking about the revolution. For Walsh, theatre seems to drag us back to a dramaturgy of finitude, even if political thought managed to escape it. The play concerns the post-independence situation, after Toussaint's death, in which the problem of leadership of an unprepared mass toward a thriving nation becomes acute. Walsh speaks of the "mise-en-scene of Christophe's tragic struggle to build Haiti" and writes that "the turn to theatre opens up another way to approach the Haitian Revolution and its aftermath, including the fault lines on which the new kingdom was founded."[12] He leaps on Césaire's interest in Nietzsche's famous essay on tragedy, and commences to reiterate the word "tragedy" as many as five or six times per page, including a single paragraph that uses the word four times, in phrases such as "the magnitude of the tragedy" seen through the "persona of the dying hero who witnesses the suffering he has wrought in an otherwise noble cause," "Christophe's tragedy," and "the tragedy of Toussaint." For Walsh, the death of Toussaint is both "the death of an innovative political project [and] the agony of a man buried alive in the Jura [a castle in Switzerland where Toussaint died in captivity at the hands of Napoleon]."[13]

It is easy, it seems, to use the theatre as a form that seems so readily to offer its tragic formula to simplify and reduce historical fieriness, temporal novelties, to extinguished ash. I hope, in my work on these texts, to withdraw this theatrical solution, and to suggest that the theatre can offer other, much less formulaic and predictable means by which to explore what happens when history is initiated, not finished. Kristin Ross, citing Claude Roy in her book on the Paris Commune, takes him to task for the way in which he designates the Commune a tragedy, a unity of time, place, and action ended by slaughter. She writes that she can turn his remark "on its head . . . . It is only by framing our perception of the event according to the laws of tragedy that the insurrection fulfills generic expectations. If we attend instead to the particularity of its unfolding and to the political culture that traversed it and that grew out of it, isolation and tragic affect are rapidly dispelled."[14] I do not wish to imply that the plays are not concerned with articulating and exploring the real politics of the struggles with colonial and imperial forces, the terrible challenges of creating a post-independence, postcolonial nation, the relation of leaders to the masses in whose name they act, and the serious impact of mistakes and failures throughout all these processes. I do not mean to suggest that tragedy no longer has a place in theatre, or to deny its astonishing generative legacy, including, for instance, its origin plumbing and regeneration in the work of Romeo Castellucci and Socìetas Rafaello Sanzio. In the work of that company's

Tragedia Endogonidia, tragedy is a language that is specific to theatre but, importantly for my argument here, one that is virtually unusable in its current form. Castellucci writes, disembedding tragedy from its expected genre markers, that

> tragedy's way of presenting a dramatic situation to the spectator is still the unsurpassed model for every intimate human representation. Our time and our lives are completely detached from any concept of the tragic. "Redemption," "pathos," and "ethos" are inaccessible words that have fallen into the coldest of abstractions. Turning again to tragedy doesn't mean looking back; we have to break Aeschylus' thread—not follow it. The theatre I respect, now, is a theatre of commotion. This is a tragedy of the future.[15]

The task therefore is not to reject the tragic genre altogether, but perhaps to think of the tragedy as a theatre of commotion, as in the plays here that are full of the lived reverberations of the idea of equality, of the idea that the slaves are no longer slaves and will not be slaves again, and of the expectation that these ideas are not going to be extinguished. James, in fact, said that he was inspired to write *Toussaint* by a spontaneous strike of Parisian workers in 1934. He wrote that these thousands of workers were "the stock of 1789, . . . 1792, of 1830, of 1848 and 1871,"[16] in which he saw the same ember of insurrection igniting again and again. These dates name the storming of the Bastille, the declaration of the Republic in France, the July Revolution, the 1848 Revolution, and the Paris Commune.

C. L. R. James, Communism, and Black International Radicalism

C. L. R. James was born in Trinidad in 1901 and became one of the foremost intellectuals and writers on communism, internationalism, anti-colonialism, and Pan-Africanism in the twentieth century. The revolutionary and black international world context into which he was born, grew up, and eventually helped to shape was galvanizing. Here is just a sampling.

In 1887 Henry Sylvester Williams formed an African Association which convened the first Pan-African Congress in 1900. In 1903 W. E. B. Du Bois's *The Souls of Black Folk* was published in the United States. In 1905, in Russia, Trotsky offered his theory of the Permanent Revolution. In 1912 Jamaican Claude McKay, a forerunner of the Harlem Renaissance, came to the United States and then went to London in 1919. He later went to Russia for the Fourth Congress of the Communist Internationale. In 1917, the year of the Soviet Revolution, the worst race riot in American history took place in East St. Louis, with white steel workers attacking black workers who both Du Bois and James called fugitive. In 1919 Marcus Garvey and the Universal Negro Improvement Association's Black Star Line began crossing the Atlantic, putting "in motion his twentieth-century spectacle of a modern black nationality."[17] In 1919 Communist Cyril Briggs founded the African Blood Brotherhood for African Liberation and Redemption, and in 1919 the first Pan-African Congress was held in Paris with Du Bois at the helm, with a reformist agenda toward colonialism. By the Third Pan-African Congress in 1923, Du Bois started moving to the left, and by the Fourth in 1927, the Congress "deepened critique of colonialism and global black oppression."[18] In 1924 Lenin died and Trotsky was dislodged from power by Stalin. In 1930 the Comintern "formed another Negro commission: the International Trade Union Committee of Negro Workers." The Sixth World Congress of the Comintern in Moscow in 1928 passed

> a series of resolutions which held that blacks living in the US South and in South Africa constituted oppressed nations within a nation . . . [and] that these groups possessed an inherent right to self-determination . . . it . . . gave an official imprimatur to efforts by black radicals of the period to search for the roots of revolutionary traditions within cultural and religious movements such as Voodoo and The Watchtower Movement, which at the time were largely seen by European radicals and liberal alike as expressions of savagery or superstition.[19]

Between 1923 and 1934, McKay was traveling up and down the French Coast, talking to seamen, "a small, transient international community of black men from Senegal, South Africa, Dahomey, Morocco, the West Indies and the United States."[20]

James moved to England in 1933, where he was reunited with his boyhood friend George Padmore. Padmore was the "leading black figure in the international Communist Movement."[21] He lived for two years in Lancashire, where he was in the midst of a militant working class, and then moved to London in 1933, where he became central to the Trotskyists. After Stalin forced Trotsky into exile, it was key for James and those he worked with, as Grimshaw writes,

> to document Stalin's betrayal of the fundamental revolutionary principles upon which the Soviet Union had been founded. *World Revolution* [James's book on the Communist International] was such an attempt. James relied largely on secondary sources, gathered from across Europe, to build a devastating case. At the core of his interpretation lay Stalin's 1924 pronouncement, "Socialism in One Country"; for at a stroke the international character of the revolutionary movement was undermined and the fate of the fragile new Workers' State was severed from the organisation of the socialist revolution in other parts of the world.[22]

James's internationalism was in the context of what Martiniquan writer, philosopher, teacher, and political commentator Jane Nardal called "Internationalism Noir."[23] In 1935 Mussolini invaded Ethiopia, one of the few states in Africa that was not colonized, and Emperor Haile Selassie was forced into exile. One of the responses was the beginning of the Rastafari movement. James responded by founding the International African Friends of Ethiopia (IAFE) in London, which became the International African Service Bureau, led by James, Padmore, and others. The IAFE was "the hub of a global protest movement."[24] But James was, according to Grimshaw, "forced to confront the equivocation of the British labor movement in the face of imperialist aggression in Africa."[25] In an essay from 1936 he wrote about the need for "an independent movement of Africans and people of African descent in the struggle for freedom."[26] Also in 1935 Du Bois published *Black Reconstruction*, which would influence James's writings on American black people and the concept of fugitivity. In 1938 he published *History of Slave Revolt*, later retitled as *The History of Pan-African Revolt* (1969) which would become profoundly subversive as a "stinging indictment of colonialism," and in which he wrote, "the great revolutionaries of the modern world needed the Africans as much as the Africans needed them."[27] And, speaking of the play that is the subject of this chapter, Grimshaw writes, "[James] hoped to make his audience aware that the colonial populations were not dependent upon leadership from Europe in their struggle for freedom, but that they already had a revolutionary tradition of their own."[28]

James, who studied and thought about the Haitian revolution extensively, finished this edition of *Toussaint Louverture: The Story of the Only Successful Slave Revolt in History* in 1934. It was performed in 1936 at Westminster Theatre in London, in a version which made cuts to the original text and starred Paul Robeson as Toussaint. (The play was rewritten in 1967, in a *much* altered version in terms of content, politics, and form, which James called *The Black Jacobins*, and which is available in *The C. L. R. James Reader*.) Two years later, in 1938, he finished his still definitive history of the Haitian revolution, *The Black Jacobins*. Nick Nesbitt provides an excellent description of Black Jacobinism, in a way that points to what is an important core of James's political thought and his portrait in the play of Toussaint's original political vision—one of federation with France, a guaranteed end to slavery, full citizenship, and a mutually beneficial coalition, rather than independence and nation.

The concept of federalism as an alternative arrangement to colonialism, as opposed to the nation-state, is one that Césaire also finds compelling in Toussaint and one that Césaire, as a Communist representative from Martinique in the French Assembly, attempted to embody. Nesbitt writes,

> Black Jacobinism, like French Jacobinism, has repeatedly named a political alliance between leftist, radical Enlightenment intellectuals (whether lawyers like Robespierre, devotees of Guillaume Raynal and Denis Diderot like Louverture, poet-intellectuals like Césaire or a priest such as Jean-Bertrand Aristide) and a mass population (of sansculottes, former slaves, or Haitians casting off the Francois/Jean-Claude Duvlaier regime after 1986) struggling for popular sovereignty.[29]

Key to me are the indications that James writes into *Toussaint Louverture* of the vivid sense of exchange and co-creation with the radical leaders of the French Revolution, before the bourgeois ascendancy, when France and Haiti were each sites for the initiation of equality and liberty. To declare independence and make Saint Domingue a nation did not make sense in light of the possibility of an ongoing federation with a republic dedicated to liberty and equality for all, a republic with intellectual and cultural resources radically open to black participation. At the same time, James continued to argue against the centrality of Europe as a revolutionary site. It is important to note that James wrote both the play and *Black Jacobins* "while he was an active member of the Trotskyist movement." For Grimshaw,

> His analysis was deeply marked by his particular political allegiance, although a number of ideas central to his interpretation of the 1791 slave revolution raised, implicitly, a challenge to certain assumptions on the revolutionary Left. He cast doubt on the assumption that the revolution would take place first in Europe, in the advanced capitalist countries, and that this would act as a model and catalyst for the later upheavals in the underdeveloped world. Secondly, there were clear indications that the lack of specially-trained leaders, a vanguard, did not hold back the movement of the San Domingo revolution.[30]

As such, he remarks that the slaves on sugar plantations in Saint Domingue in the late 1700s "were closer to a modern proletariat than any group of workers in existence at that time."[31]

I stress the internationalism of the situation, its communism, its innovations in political thought, and its insistence on the black revolutionary artist and intellectual as central to modern history and politics. If I am writing against an evaluation based on the tragic, and hoping to impress upon my readers the political novelty and potential of the temporal suggestion of the new present, it is important to emphasize the constellation of radical thought and practice, the "transnational black public sphere,"[32] as Gary Wilder calls it: the newness of political ideas, the activism, and the commitment that was borne on anything but a tragic sense of history.

Plays on the Haitian Revolution

There are three plays on the revolution that would make for a fascinating event done as a trilogy and that I originally intended to study as a trilogy in this book. While I will very briefly describe the third here, I have chosen in the interest of space to focus only on two. In the case of *Toussaint Louverture*, I have chosen to include a few very brief segments in which I share a kind of directorial imaginary for staging it. Unlike anywhere else in this book, I am practising in this chapter a close textual analysis, which begs the question of whether I mean that the text itself yields the new present, or whether it depends upon a staging, a collaboration with a production team, that differs from the playwright's specifications. In the case of *Toussaint*, which is perhaps less stage-worthy than *Christophe*, I have decided upon these brief excursions into mostly unrealizable staging: a method for a partial extension of my ideas into the potential of the lived world of the performance space. These descriptions are written informally to indicate an open-ended, indecisive exploratory sequence.

Toussaint Louverture: The Story of the Only Successful Slave Revolt in History is a study of the galvanizing leader of the Haitian revolution that initiated the emancipation of the slaves whose labor fed the insatiable

European appetite for sugar. When it started there were 500,000 slaves in Haiti, along with 30,000 whites and 30,000 mulattos. There were 800 sugar plantations, which created the most profitable trade in the Antilles, part of the circum-Atlantic development of capitalism based in part on plantation slavery. Nantes, France was, as James writes, "the centre of the slave-trade [and] as early as 1666, 108 ships went to the coast of Guinea and took on board 37,430 slaves, to a total value of more than 37 million, giving the Nantes bourgeoisie 15 to 20 per cent on their money."[33] Importation of slaves grew and grew, so that by 1787 40,000 slaves per year were being brought to Saint Domingue (the colonial name for what became Haiti), "the most profitable colony the world had ever known."[34] The importation of so many slaves from Africa, "used to open up new lands," meant even more brutal methods were used so that they could "be broken and terrorized into labor and submission."[35]

The play begins with revolution's initiation in its brutal colonial context. Toussaint was a slave, but a literate coachman, married and with children; two years later, in the play, in his mid-life, he assumes the leadership of the revolution as a man who would come to prove a brilliant intellect and military strategist. The play is epic in structure, situated in many locations, across a wide swath of time, and ends just after Toussaint's death in 1803. The play follows the strategic twists and turns he and the revolutionaries make in order to remain true to the absolute declaration that there is not and will not ever again be slavery on the island. Two men fought fiercely alongside Toussaint as fellow leaders during the slave revolt. One was Dessalines, who became the leader of Saint Domingue following Toussaint's arrest and removal. His armies successfully fought the French invasion launched by Napoleon intended to reinstate slavery, and he declared Saint Domingue an independent nation in 1803. He renamed the island Haiti, its Arawak/Taino name.

The second fellow leader, Christophe, is the subject of the second play, *The Tragedy of King Christophe: A Play*. Like the first, it follows a narrative path, in this case tracing the downfall of Christophe from his post-Dessalines crowning in the midst of a civil war with the mulatto leader of the southern region of the new country, Pétion. The play moves through his driving and destructive will to establish Haiti as the equal of any European nation by the cruel enforcement of impossible work regimes and his retention of a land policy in which the elite black and former colonialists still had most of the land, to the uprising against him and his suicide. The second play has a multiplicity of registers, including in its poetic, mythic, and linguistic structure. The first is also stylistically inconsistent, in conventional theatrical terms, in its use of often-clashing forms of what we might call genres. In their overall structure, however, both plays are epic, in that they are multi-scenic, covering vast geographical and temporal differences even from scene to scene.

Edward Glissant's *Monsieur Toussaint* is the third play. It returns to Toussaint himself, dying alone in his prison in the Alps with his single caretaker, who is sometimes joined by others of the French "living"—the commander of the fort in which he is imprisoned, an envoy from Napoleon. It is a four-act play, with each act consisting of short scenes during which the dead, specters of his own imagination, melt into the cell and out again. He is visited, plagued, remonstrated by his fellow revolutionaries, all of whom are dead, one even at Toussaint's own hands. One of these Haitian figures, Mackandal, was the leader of a group of fugitive slaves, the maroons, who enacted guerilla warfare against the plantation owners in 1756, decades before the 1791 meeting in the woods that formally initiated the beginning of the Haitian revolution and begins James's play. Although Mackandal does not appear in either of the first two plays, it is worth saying a little more about him since he is deeply ingrained in the scene of the 1791 revolution. Mirzoeff describes him, (one of his arms had been cut off), as "a disabled, Islamic, sexually promiscuous, African resistance leader and priest in the plantation economy."[36] He was born in Africa and became a slave when he was 12. He both read and wrote Arabic. Forming successful networks of insurrection across Haiti, he came to embody a "garde-corps" (body guard). These could "induce forces from the dead to protect the living."

The garde-corps had the capacity for second sight, and were in fact called upon by summoning this sight when needed, in the case of Haitian revolts, to change "the balance of power in the colony."[37] Thus, Mirzoeff contrasts Mackandal's second sight, which he endowed all the insurrectionary slaves with, to "European oversight."[38] The colonists' first attempt to burn Mackandal at the stake failed. Somehow he managed to free himself and escape. He was recaptured and subsequently died at the stake, but "his revolutionary and magical powers remain part of the cultural memory of Haiti to this day,"[39] as if he himself, as I will argue about Toussaint, is a component of possible new presents.

Toussaint Louverture: The Story of the Only Successful Slave Revolt in History

The play opens with a scene that is an illustration of the plantation owner Bullet's brutal treatment of slaves and absolute rejection of the French revolutionary government's petition that the mulattos on Haiti be declared as having equal rights to whites. In his fury, Bullet has the mulatto emissary making this proposition hung on the spot and a search undertaken for the man's slave, who slipped away. Bullet orders that until the slaves find him, one slave a day will dig his own grave and be buried in it up to his neck with his face covered in honey and molasses. James gives us the political situation; while the opening party on the veranda proceeds, scored with *Don Giovanni* (from the contemporary European marvel Mozart), flush with ladies in party dresses, Bullet reassures himself that blacks will never "strike for liberty" because "such abstract notions do not enter their heads."[40] But all through the scene there are drums, voodoo drums, beating from the distance. The presence of these drums from the very beginning, puncturing the European music and white brutality, signal the existence of those who for Bullet are inexistent, in the sense of being without political or revolutionary capacity. Referring to the drumming, he says, "Listen. They are quite happy; dancing in the forest somewhere with their drums and making their heathen sacrifices."[41] Mishearing the import of the drums, believing the slaves incapable of initiating revolution against the planters to free themselves, Bullet cites the economic dependence of the new French government on the highly profitable sugar plantations to assure the planters that France will never free the slaves.

Scene 2, which follows historical accounts, is where the play really starts, where what I will call the historical riot, following Badiou, intensifies, localizes, and becomes organized: August 6, 1791. It is a torch-lit forest scene, the soundscape is constant drumming; we must hear the polyrhythmic drums of Haiti as we listen. It is a kind of revolutionary assembly of slaves in a voodoo key. The second and third lines of the stage directions are "they, the Negro slaves, are the most important characters in the play. Toussaint did not make the revolt. It was the revolt that made Toussaint."[42] The idea of equality that had flared in prior episodes of what Badiou might call the immediate riot, in Mackandal's uprisings for instance, are here organized into a dangerous, effective, and inspired revolutionary mass movement, by those who James, as I have already said, thought were "closer to a modern proletariat than any group of workers in existence at that time."[43] It is the onset of a new present.

I must pause here to articulate Badiou's temporal innovation in more detail. The new present is a radical temporal theory that there is no history. There are only historically distributed presents. There is of course history, and Badiou of course speaks of it. But what is meant here is that what is important to an emancipatory imaginary is to attend to former presents in which a particular emancipatory truth or idea appeared. The past, history, is here articulated as the appearance of these presents in subsequent new worlds, as new presents. This is the case even though the truth or idea that arrives with the new present has been extinguished in prior worlds. Each new present is specific, specifically situated and conditioned, even as the idea it carries retains its singular power. Each time it appears, the idea—equality, for instance—retains its axiomatic assertion, even as it, or the work that must be done by it and for it, is different in each case. What is important is that we give our attention to these glimpses of something that will not disappear no matter how long it may be, in another of Badiou's phrases, dragged under the bar of occultation. This is our own emancipatory capacities, lived in particular ways in the materialities of particular worlds in which they have flared again. In a world in which so much disillusionment of radical political hopes encourages repeated analysis that assumes the failure and consequent disappearance of any radical initiatives, it is the suggestion of a form of political temporal thought with which to identify, instead, that initiatives do not end. The same internal axioms (such as there are no more slaves) appear again. It is as true of the Ferguson riots, the Black Lives Matter movement, the Black Panthers, as it is of the Jacobin segment of the French Revolution, the Haitian revolution or the Paris Commune. The flare will diminish in intensity, but it will have been articulated and practised by the body galvanized to make a change to a particular world, a change which will have, as the Haitian revolution did, a universal or global address.

The slaves of the Haitian revolution, of whom Toussaint is one, initiate on that day in 1791 what will emerge into the world with vivid intensity as an axiom that there will be no more slaves. This upsurge will have, in terms of the consequences it leaves behind, a duration, a very long duration—ten years, in fact—through the army of slaves, which we will call, in Badiou's vocabulary, a "body." When he says body, Badiou is deliberately wrenching the word away from its biological, humanitarian, "democratic materialist" connotations and instead using it to name that being which forms to explore the consequences of what has appeared at critical nodes, "points," where there needs to be a decision, a decision that will determine how effectively the consequence or trace of the appearance of the idea can be adhered to.

It is clear in James's play what the body is. It is the slaves, along with their apparatus of destruction and revolt, especially fire, the forests, the geography, and even diseases, especially yellow fever, of Haiti (which foil the Europeans). This body is the "worldly element of changing the world."[44] When Badiou discusses the historical riot, he uses the word "organization" for this function, the place where "evental power [transforms] into temporality." He writes, and I think we can take this as another way to understand the new present, that the organization/body is "*the invention of a* time . . . a time that in a sense unfolds its beginning. This time can then be regarded as an *outside time*, in the sense that organization is not amenable to being inscribed in the order of time dictated by the previous world."[45] It is clear that in that meeting in the forest there is an invention of a time, cadences and fugitivity and multi-rhythmic musicality, that "is not amenable to being inscribed in the order of time dictated by the previous world," which is colonial time, time assured, time unable to conceive of its interruption, accompanied by the stately European time signatures of Mozart.

In Scene 2, that forest scene, fire-lit, during which, in James's stage directions "the rebellion was decided upon,"[46] Boukman, a leader of slave revolts in Haiti throughout the preceding decade, begins the already fevered rally by saying "Liberty—Equality—Fraternity. The white slaves in France—they suffered like us—they've made a revolution. They killed the slave-owners—made every-body free. They divided up the property, and now in France they have liberty, equality and fraternity (*the crowd cheers*)."[47] It is important to stress, again, that this night in 1791 is the culmination of flares of insurrection that, according to James, were begun "a hundred years before 1789."[48] While, as I have said, Mackandal does not appear in the play, in *Black Jacobins* James does describe the much-feared fugitive slaves who lived in the forests, free, slaves no more. By 1751, James says, there were 3,000 maroons, living in separate groups, but often joining together under a single leader to raid and burn plantations. He cites Mackandal as the most important of these, the one who "conceived the bold design of uniting all the Negroes and driving the whites out of the colony."[49] Boukman, also a maroon and a voodoo priest, comes from this pre-revolutionary insurrectionary movement.

Toussaint, on the other hand, was a more privileged slave, with some education; he was literate, a Catholic, and, as I have mentioned, a coachman for a plantation owner. Initially in the play he is the voice of moderation. He admonishes Boukman and Dessalines to wait, to try a petition. Dessalines, as I have also already mentioned, will become Toussaint's general and many years later, after Toussaint's death, will be the one to declare independence from France and to rename Saint Domingue Haiti. In opposition to Toussaint's suggestions, Dessalines climbs up onto a platform and, rousing the crowd (and we must not forget to hear the drums, a constant in the scene), declares "no more work, no more whip. If we kill the whites we are free." James indicates that the drums are quickening with Dessalines's speech. We should hear the excitement, the eruption, the (revolutionary) fervor as Dessalines says, "I, Dessalines, will work no more. Liberty!" And the gathered slaves echo with "Liberty! Liberty!"[50] As the meeting continues, in a full fervor of commitment by the gathered slaves, Boukman delivers the speech known as "Boukman's prayer."[51] At its culmination Boukman rips the cross from his neck, followed by others, and, with the drums very loud, they drink from a bowl of blood. The reluctant Toussaint drinks too, and the slaves at the end of the scene, according to James's stage directions, "melt away in different directions," now guerilla fighters, insurgents, a new army, saying "Liberty or Death! . . . the password with which they bid each other farewell."[52] The scene brings the French Revolution into proximity with this voodoo site of the initiation of revolution, as well as implicitly staging previous decades of slave uprisings. Thus, very early in the play, James puts into place the building blocks for that eventual vision of Toussaint's in which, as I have said, James is manifestly interested: a transnational vision of federation with a revolutionary French Republic and a colony of free black citizens, formerly the Black Jacobins, who "were to make history which would alter the fate of millions of men and shift the economic currents of three continents."[53]

I keep seeing the sides of the stage as contained by virtual walls, almost not present, alternatively present and less so, not especially visible, not solid, shimmering. Maybe the stage itself is all flame at opening image. Or, maybe, the play begins with fires on mountains—signals—the symbols of revolt that came before 1791—a rolling ball of fire from the last into an explosion in that woods—a small little unit set is rolled forward—it's got Bullet's party scene—it is sort of in the middle of this ball of fire. The little unit set somehow crumples in place—melts—into Scene 2 in the forest. The theatre is filled with the smell of burning sugar. The European and colonial characters belched up from a fracture—maybe puppets or maybe human, but they lie around (the fracture gradually closes up). Sugar cane grows from their bodies. Maybe each is guarded by a slave who is no longer a slave. When it is their turn to "act," their bodies are perhaps set up, propped up, lifted to standing, and from their bodies extend small stages, or scenes. These are points, a constellation of these around the stage, in this burning San Domingue. Each time Toussaint comes, he comes with an army. Unlike in the play as written, the slaves who are no longer slaves are present—somehow, visually, they are holding the new present in place. Perhaps there is a sense that everything gets brought to, or is already in, this space. Exits and entrances perhaps only by the revolutionaries. There are drummers high up and all around. Their drumming can be in part the multi-temporal Haitian drumming, but also something that is specifically composed with an ear to new tempos, and the noise of the crowds created according to what the new present sounds like. They are in scorched clothing. The space of the stage is regulated, perhaps by these figures. Everything is brought to this space, everything that needs decided upon.

In Scene 3, set two years later in 1793, years during which plantations have been burned down, land and goods seized from the owners, Dessalines delivers his axiomatic assertion to Toussaint: "There are no more slaves." From the legal point of view this was patently untrue, since the National Assembly in Paris did not abolish slavery in the colonies until 1794. But for the slaves, for Dessalines, the truth of the situation, in this new present, is that "there are no more slaves. We are all free men."[54] Right here, now, the slaves are not slaves. This is an idea. It is the idea of Spartacus, as Badiou suggests and I expand on below, explored by the revolting slaves of that particular Roman world, an idea brutally obscured. But trans-worlds, here it is in the very different conditions of a colonial plantation culture, again opening the possibilities for emancipation. Sylvia Frey writes about an incident in 1789 that happened two weeks before the French Caribbean heard word of the storming of the Bastille, in which the military commander of Martinique received a letter written by two Martiniquan slaves. It was signed "us blacks." Frey writes that "the letters departed radically from the standard abolitionist advocacy of gradual emancipation and embraced violent self-liberation instead." Her quote from the letter is worth citing here because it is another iteration of the idea: "We the Negroes . . . are ready to die for this freedom, for we want to and will obtain it at any price."[55]

I should say before I continue that in Badiou "Idea" is capitalized. Many people misinterpret its upper case designation as indicating a neo-Platonic transcendent. It does refer to axioms that can be practised in differing worlds, but it is practised materially and complexly and is not an *a priori* transcendent. Nevertheless, to make the work here perhaps a little less abrasive to some readers, I myself will use the lower case when I am working at applying the concept. And so, the idea, that fullness of equality, declared axiomatically in "there are no more slaves," existed in fascinating echoes, in varying degrees of intensity, in relation to the French situation, as Boukman's initial speech to the slaves makes clear. We could also borrow the phrase from Nick Nesbitt that I have already cited—"the militant defense of the idea of undivided equality"—to describe this idea. Nesbitt describes it as belonging to Jacobinism, both French and Black, each of which sought to "implement a dictatorship of the masses in defense of popular sovereignty."[56] For Badiou, the idea must be applicable to everyone. As such, a slogan which belongs to an identity politic, a racism, a neocolonialism, a fascism, a neoliberal humanitarian practice, cannot be an idea. It is a new vision of a universal generated here, in Haiti, and it is to the universal that we must accommodate ourselves if we desire emancipation. As Gary Wilder writes, "Understandable fears of totalizing explanation and Eurocentric evaluation have led a generation of scholars to insist on the singularity of black, African, and non-Western forms of thought. But we now need to be less concerned with unmasking universalisms as covert European particularisms than with challenging the assumption that the universal is European property."[57]

Nesbitt, in his book *Caribbean Critique*, takes Badiou to task for linking the Roman Spartacus to Toussaint. The figure of the black Spartacus is taken from a man who James says was the only person Toussaint trusted, his only friend: the French General Etienne Laveaux. Laveaux drew from a book by the French priest Abbé Raynal, *Philosophical and Political History of the Establishments and Commerce of the Europeans in the Two Indies*. It was famous in its time. According to Susan Buck-Morss, it was "widely read and not only in Europe."[58] By 1780 Raynal was calling for a slave revolution and a black Spartacus. Laveaux, naming Toussaint assistant to the governor, called him "the black Spartacus, the Negro predicted by Raynal who would avenge the outrages done to his race."[59]

Nesbitt unfortunately misreads Badiou in two important ways. Badiou writes that the axiomatic assertion of the Roman slaves was "we want to and can return home." For Badiou, this declaration is "indexed to the possible,"[60] and indexed to "the present of a hitherto unknown possibility."[61] Nesbitt cites the phrase as the slave who "wants to . . . return home."[62] By virtue of this curious elision, Nesbitt implies that Badiou is literally referencing the Spartacus slaves' desire to return home, as if Badiou's were a vision of domesticity. He misreads the fact that Badiou is describing the initiation of a new present when the slaves say, "I want to and I *can*." The "can" is an axiomatic declaration that means I am no longer a slave, now, in this new present. Nesbitt claims that Badiou misses the fact that the Saint Domingue slaves could never return to their homes in Africa and therefore that he is wrongfully applying the thought of "a European [Laveaux] unable to grasp the world-historical event unfolding before him." For Nesbitt, Badiou, in speaking of "home," is defining the Spartacus revolt "as an archetypical conservative revolution that failed to envision the destruction of a world in its entirety, one that remained content to envision merely the escape of individuals from that system to an individualistic freedom." Badiou is incorrect in correlating the two slave revolts because, according to Nesbitt, he misses the fact that Toussaint "radicalized the gesture of innumerable slave revolts throughout history by producing a revolution that destroyed the world of plantation slavery in its entirety."[63] We have only to check with Badiou, in the same section of the same book Nesbitt is working from, to find a statement that directly contradicts Nesbitt's interpretation. Badiou writes:

> This is the revolt which made the principle of the abolition of slavery real, which conferred upon blacks the status of citizens, and which, in the exhilarating context of the French Revolution, created the first state led by former black slaves. In sum, the revolution that fully freed the black slaves of Santo Domingo [*sic*] constitutes a new present for the maxim of emancipation that motivates Spartacus's comrades: "The slaves want to and can, through their own movement, decide to be free." And this time, the white owners will be unable to reestablish their power.[64]

In making this curious mistake, Nesbitt is able to criticize and extract the European thought that was so productively important for Toussaint himself from the Haitian situation. Such a move, and others like it, deletes the promising possibility for *temporal* thought that Badiou, European as he is, introduces. Badiou, in the affiliation he creates based on Raynal and Laveaux, is able to suggest a global consideration of various new presents, a temporal philosophy of revolt, which Nesbitt's move to eliminate the European from consideration prevents, at least in this passage of his book.

Nesbitt's opinion that Badiou considers the Spartacus revolt a "conservative revolution," a failure "in its entirety," is the second of his misreadings. Badiou does indeed mention the brutal suppression of the Spartacus revolt. But he does not speak about it as a failure. What he does do at this point in the text is to unfold the theory of the new present that *accounts for* the ways in which aspects of an insurgent event, a coming into existence of the inexistent in full political capacity, are folded back into a regrouping by the powerful, while the truth that appeared in that event moves into only temporary dormancy. The forces of reaction and obscuration respond to the new conditions of a world brought into visibility by the event, the revolt, the insurgency, which caused the "transformation in the rules of visibility themselves."[65] But a truth "created in history" cannot "slip back into nothingness."[66] He writes, "of no truth can it be said, under the pretext that its historical world has disintegrated, that it is lost forever."[67] These are his terms by means of which the Spartacus and the Haitian are compared. To study the forces of reaction in the Spartacus revolt is not to say that "it failed in its entirety," but to be pragmatic about the subtleties and degrees by which a truth—there are no more slaves—will be temporarily re-submerged.

I have already spent some time on Badiou, and I will try to save my reader an account of the full scope of Badiou's analysis here in the interest of economy and returning to the plays at hand. But I must for the sake of my argument draw my reader a little further into his ideas. And I cite him repeatedly because the exacting precision of his language is hard to paraphrase accurately. I have already talked about the body that forms around the appearance of the idea and explores how to go about realizing it in actuality. The body comes up against points of decision, political and military strategy, the prospect of brutal punishment, and so on. But the body is divided—not all of its parts are equally at work on this new present, and some parts of the body will make choices in contradiction to other parts. The army is forming, in formation, under the sign of the trace, under the bar of the trace, "something like the active unconscious of the trace of the event." Some of the army are what Badiou calls "faithful subjects,"[68] because they actively produce consequences from the event, or the appearance of the idea. In that first revolutionary scene of the play, we can say that the slaves who slip away in James's stage directions have taken upon themselves the role of faithful subjects to Boukman's axiomatic declaration of liberty.

But there are also slaves, in Badiou's example of the Spartacus revolt, who did not join the body. These, who he calls the reactionary or conservative subject, cannot exactly "negate" the consequences extending from the appearance of the idea. This is because this subject itself "claims to produce some kind of present." That is, this subject, reacting to the new, comes up with a "measured present . . . a present a little less worse than the past." This reactive subject, for whom the appearance of the idea seemed "a catastrophic temptation," will take comfort in the minor improvements made in response to it. In this way, the reactive subject can experience a feeling of belonging to something new, even when the new present has turned into the "extinguished present."[69] In one of my favorite phrases—favorite because it rings so true—the reactive subject, even as he is "perturbed by the insurrection of the friends of Spartacus," belongs to "a lustreless form of the present."[70]

The third type of subject is the obscure subject. Unlike the reactive/conservative subject who hones to a lusterless, diminished slice of the present, the obscure subject considers the entire new present "malevolent . . . inexistent." The entire new present is placed under the bar. (Badiou illustrates this set of ideas with formulas similar to mathematical formulas, and so the bar is the mark of what is put below, deleted.) We can use as example here Napoleon Bonaparte. As with the reactive subject, the obscure subject produces the "night of the present" but, while the obscure subject desires the full return of the old order, its obscure operation is nevertheless contemporary. The form of obscuration must respond to the new conditions created by the faithful subjects, the "rebel body."[71] The obscure body calls upon "an atemporal fetish: the incorruptible and indivisible over-body, be it City, God or Race."[72] But, even though reactive and obscure subjects immediately intervene in the new present, Badiou insists that "a fragment of truth inserted under the bar by the machinery of the obscure can be extracted from it at any instant."[73] That extraction is called the new present.

And so we can return to this this phrase of Dessalines: "There are no more slaves. We are all free men." And a vibrantly revolutionary scene, at the end of Act I, takes up the French site of this idea. The scene is a play within the play and is, interestingly, a verbatim theatre piece, inscribed by James word for word from the transcripts of the 1794 decision of the National Convention in Paris to abolish slavery. It is staged in France. A mulatto and a black man from Haiti are invited into the assembly and embraced as representatives of those "who have suffered oppression for so many years."[74] The declaration freeing the slaves is given, and the hall resounds with "*shouts*" of "long live liberty."[75] In the scene back in Haiti, interrupted by the insertion of the Convention, Roume, a member of the French revolutionary government, says to Toussaint:

> In Europe the ancient monarchies are attacking us on all our borders, determined to crush this first great uprising in the whole history of mankind. More than our bullets and guns they fear the words liberty, equality, and fraternity, which tear the veils of tradition from the minds of all the oppressed who hear them . . . . But the revolution will triumph, if all the sons of liberty, white, black and brown fight for it wherever they are.[76]

Thus James lingers on and enforces the sustaining affiliation between both French and Black Jacobinism, even as it is clear that the French site of the declaration of "the militant defense of the idea of undivided equality"[77] is conditioned by the Saint Domingue site. Of course the slaves did not need Europe to teach them about equality, to teach them revolution.

> Maybe for the scene in France during which slavery is abolished, the stage is swept entirely clean and empty—a clearing. And the Convention scene is staged in full. When the scene is over, somehow we are back to Haiti, but one that looks different—a magical effect where the real of the performance in France is kind of sucked back into a corner—like being sucked into the funnel of a tornado—where it reappears on a pedestal in model form. (Maybe the pedestals are twisted wreckage—maybe some are almost horizontal—a kind of landscape of the new present which holds these points and this ruin of its appearance.) This new Haiti environment.

I want to return now to Act I, Scene 3 in more detail, the scene I have already used to lead us to Badiou's conception of the idea in Dessalines's axiomatic assertion. It is this that persists through this scene, which is complicated and difficult. James attempts to do a great deal here, including the parsing of the need for and character of a leader for the revolutionaries whom he at times portrays as already treacherous to one another, even ignorant, while at the same time preserving a sense of their absolute fidelity to the end of slavery. It is 1793. The "leaders of the rebellion" are dressed in French epaulettes and bandoliers, sitting at a set of Louis XIV table and chairs, all the proceeds of their plunder. It is the slaves' camp. Boukman and Dessalines are training the army in military maneuvers. In the arms of these men, according to James's stage directions, are "old muskets, cutlasses, spears. They are as unkempt as they were on the night when the rebellion was decided upon, but they carry themselves with new confidence."[78] The parodically ridiculous and grandiose character of Macoya joins them, coming from the Spanish part of the island, San Domingo, with an offer of aid from the King of Spain. Macoya argues for loyalty to the institution of monarchy against the French revolutionaries, among who he groups the colonists. The King of Spain, he says, will support the slave revolution since he considers them to be fighting the anti-monarchical colonists just as he is fighting the anti-monarchical revolutionaries in France. Macoya reports that the King of Spain will make them all generals in the Spanish army.

Toussaint, the only one who can read and write (and is thus promoted to the position of Boukman's secretary), announces, in the midst of the slaves' debate about whether to accept the offer, that he has invited a representative of the French revolutionaries, Roume, to come to meet with them. Toussaint says that they should wait to see, that he may be offering the slaves terms. As it turns out, Toussaint has already drawn up a list of terms to present to Roume. If he will agree to them, this will mean, for Toussaint, that the slaves will stop the rebellion. They are moderate terms, "stagist" terms, revisionist terms, not revolutionary ones. Drums sound from the hills, signal drums, drummers set up by Toussaint to announce the passage of Roume into their camp. As they wait, Dessalines systematically rejects each of Toussaint's terms and this is when he delivers his axiomatic assertion to Toussaint: "There are no more slaves. We are all free men." James consistently portrays Dessalines as the loose cannon, the one who may menace revolutionary strategy with his impulses toward vengeance, the one who wishes to achieve victory by driving the whites into the sea. James is beginning to build here a sense of the divided body and of decisions at points—those decisions that will determine how the slaves can or cannot continue to act in fidelity to the consequences of the revolt initiated on that night in August 1791. And yet, while another reading might find here in Dessalines and others of the leaders present a portrait of the immoderate, short-sighted, and violent components of revolutions (which will come to a tragic end thereby), I would suggest that James nevertheless, and despite his own suspicion of Dessalines, preserves that axiomatic, from which not a one of these revolutionaries retreats.[79]

When Roume enters, he is accompanied by Bullet (from Scene 1), who is now a leader of the Colonial Assembly. Presented with Toussaint's demands, Roume is—according to the stage directions—surprised at how moderate they are and seems to agree to them, pending approval in Paris. However, Bullet intercedes, distinguishing between the revolutionaries in France willing to entertain the idea of "political rights for free blacks" and the planters, telling the revolutionaries to "come back like the dogs that you are and beg our pardon . . . . You are the property of the planters, bought and paid for and keep that always in your black skulls. The Assembly might pardon you if you repent and get back to work."[80] He tells them that the Spanish are using them against France and will throw them back into the fields when they are done with them. And he informs them that the king of France has been beheaded. Bullet here is directly obscurantist, drawing the revolutionary situation back under the bar of the old order in the form of even more vicious racist violence—culminating in whipping his former slave, Jean-Francois, now a revolutionary, in the face.

The scene takes a strange turn here—strange because three of the leaders, Jeannot, Jean-Francois, and Boukman, are horrified by the news of the beheading and declare to Roume that they cannot agree to terms with him since he represents a people who have killed their king. It is strange because in Scene 2, set two years before but in production probably less than ten minutes earlier, Boukman, you will remember, gave the fiery speech about the white slaves of France who killed the slave owners and distributed the property. And yet here, the news undoes any possibility of making a deal with Roume, even as he tries hard to convince them that it is the revolutionaries, not the former king, who are on their sides. James has Jean-Francois specifically say: "From the beginning of the world the blacks have executed only the will of a king."[81] Toussaint is quiet through the exchange, presumably watching and evaluating, presumably considering the acute violence with which Bullet acts and deciding against his own gradualist approach. The others eventually turn to ask him what he thinks. He says, "We cannot join those who have killed a king."[82] What remains confusing is why James has his revolutionaries wedded to kings when he had them initiate history through a cry of solidarity with the white slaves in France. James writes in *The Black Jacobins* that "like all other blacks, Toussaint attacked the godless kingless republic and fought in the name of royalty, both Spanish and French. But for him, already, these slogans were merely politics, not convictions."[83] James goes on to remark that many scholars have believed that Toussaint "had some 'African' faith in kingship" but that this is wrong—James, deciding to make war on the colonists aided by the Spaniards, used the "prestige" of being in a king's army to continue "to rally the blacks on the slogan of liberty for all."[84]

So perhaps we could say that in writing the scene in this way, James portrays the revolutionaries as members of the body divided, here by traditionalist institutional affiliations operating in an entirely new situation. In this way he can set up a point, a choice Toussaint must make, which is to pretend momentarily to this affiliation himself. This way he can, under cover, reunite the body at this point of decision against both the French revolutionaries and the Colonial Assembly, who he will not have failed to observe are at this point still entirely imbricated with one another. And this is exactly what he does. He acknowledges to his comrades that they were right, that they can expect nothing of the planters, cannot expect anything to be approved in the Colonial Assembly, and therefore that they will have to wage war. His own moment of transformation is at hand, as he asks to become a soldier with his own company because "the time for letter writing is over."[85] In a sense, this is a point at which, again, the new present, "not amenable to being inscribed in the order of time dictated by the previous world," "unfolds its own beginning."[86]

James is careful to set up the complexities of this revolutionary situation, which he understood to be central to Toussaint's political vision—the sense that Toussaint's fellow revolutionaries were not prepared enough, educated enough, to continue the revolution without a leader. This is of course consistent with James's reading of Toussaint as exemplifying the problematic questions of the leader versus the mass. The scene ends as everyone has left the camp and Toussaint sits alone, worried and upset. But then he is interrupted, first by Boukman and then Jean-Francois, competitors in leadership and each of whom wants Toussaint to join his side. Toussaint is non-committal with both. When they are gone, he despairs of either of them and opens his Raynal, to which I have already referred. He reads aloud, verbatim, starting from the line "a courageous chief only is wanted."[87] (In *The Black Jacobins*, James provides more of this text from Raynal in making his account there of this moment in Toussaint's life.) Toussaint says to himself, "White men see Negroes as slaves. If the Negro is to be free, he must free himself. We have courage, we have endurance, we have numbers." And in the next breath, he understands himself to be the leader of the in-process revolution, the "great man" "owed" to the oppressed: "I shall be that leader."[88]

It is important to acknowledge and retain room for the questions of revolutionary unpreparedness and leadership that James is contemplating here. And yet I think we can understand the classic question of the leader slightly differently in this scene. Toussaint has been reading and thinking about this passage in Raynal over and over. We can imagine, in the vocabulary of this chapter, that it echoes for him as part of the evental upsurge, which he has already experienced and by means of which he has already begun to belong to a body. It means the upsurge in which the inexistent (in Badiou's terms, those who do not have political existence) make their existence visible, in which slaves free themselves. It means that history starts at this moment of self-initiation in the name of that which will have a universal address; in this case, there are no more slaves. We might read in this moment the appearance of the Black Jacobin, the "tropicalization and radicalization of Jacobinism."[89] We could say that Toussaint is thus making a decision at this point, where the divided body falters, that will ensure that the revolution continues. It is a decision conditioned by James's initial stage direction, that the revolt made Toussaint, and not the other way around. That is, the condition of taking up leadership is to be understood as one that belongs to the body, within the body itself, as one of its decisions. James makes a point in this scene of referring to Toussaint by his former last name, Breda. Although James does not overtly stage it, at the moment that Toussaint made this designation of himself as the particular part of the body that is the leader, he renamed himself "Louverture," or the opening, the beginning. It is a name that signifies the leader as an opening for the body to pass through, for the revolution to open by means of, for the revolution to really begin.

The play, in order to be alive with this new temporality—which saves it from being a tragedy about failed revolution—needs to be topologically stretched, centers of vitality found. The intensely masculinist discourse of revolution here must be cut through, sliced—by something stronger than simply adding the representational presence of women to the stage. Flares of multiplicity, woman/man/indeterminate/unfinished, in all the slaves—in the soundscape, the gestures, clothes, attached objects, the elements of the body, water, forests, rain. It must depart from the masculinist imaginary that Michelle Ann Stephens so astutely and importantly describes in black revolutionary figures. Undo.

The slaves who were slaves no more burned the plantations, razed them to the ground. It needs to be topologically active, internally fiery. From Glissant's play:

> **Toussaint** Discord has been put aside, a country of slaves, purified by fire, is thriving more than ever.[90]

The battles need to be restored to the stage. The tragic genre traditionally places violence offstage, and perhaps James was following this as much as thinking about the realities of staging. Maybe the scene with all the scheming Europeans is just their recorded and amplified voices, distorted, while on stage the battles, and perhaps the simultaneous rolling out of a field and its laborers. Sugar cane appears. No process of growth per se. Indications onstage that defeat futurity-oriented thinking like "planting seeds." There is the army, the war, and there is the labor within this present.

As the play goes on, the situation presents ever more complex and demanding points of decision for this revolutionary and divided body of slaves who are slaves no longer. It is clear in the play that there are some parts of the body that are efficacious with regard to the decisions at points and others that are inert or even negative. The mulattos, for instance, are disengaged, even hostile, parts of this body, which is, therefore, divided. To review, the new present is complex, internally challenged, inevitably met with forces of reaction and obscurantism, re-engineering a night (as Napoleon works to restore slavery). But—and here is where this inevitable disappearance of the new present differs from tragedy—this night, as Badiou says, must be "produced under the entirely new conditions which are displayed in the world by the rebel body and its emblem."[91] That is, Napoleon cannot enclose the slave revolution in obscurity, into an "extinguished present," without having to deal with the entirely new conditions produced by the slaves who are free in the new present. Toussaint, in his jail cell—the scene of his death—says that "the First Consul can never restore slavery in San Domingo" because "you can defeat an army, but you cannot defeat a people in arms . . . do you think an army could drive those hundreds of thousands back into the fields?"[92] and (the following is quoted verbatim from a famous piece that Toussaint wrote) that "in destroying me you destroy only the trunk. But the tree of Negro liberty will flourish again, for its roots are many and deep."[93] In these words he explicitly names the conditions that the new present, however now obscured, have brought into being.

The scene of Toussaint's capture (Act III, Scene 3) is a scene of a kind of resurrection of republican ideals in the key of slaves who are slaves no more, as Toussaint marches to meet the French with his troops, the entire city gathering and singing "La Marseillaise." A French officer and the wife of the French General Leclerc are angered and made afraid by the growing force of the march and the music, saying that that song is the song of the republic, to which the "black savages have no right."[94] Suzanne, Madame Leclerc's mulatto "companion," says, "But it is our song too, Madame . . . when we fought for freedom we sang those songs. The revolutionaries in France sang them and we sang them too."[95] The play ends with the rallying of the slave masses once again for their again-threatened liberty in fidelity to the now-dead Toussaint, even as his former general Dessalines, in the name of "Liberty or Death," and in the name of Toussaint, declares at that moment the independence of Saint Domingue, renaming it Haiti. In this treatment of the Haitian revolution by James, something new was created in the world. Something fragmentary and brief, but that, most untragically, will come again.

In his full and detailed treatment of James and tragedy, Glick provides, as I have said, a Marxist analysis to identify the tragic structure in the play, and more fully in the 1967 version. As fascinating as the work is, Glick remains in some senses wedded to a set of perceived benefits in the genre, arguing that James was both using the tragic genre and modifying it for his revolutionary anti-colonial and anti-imperialist purposes. Glick writes that for James the tragic genre was useful because of "its root as a key ritual of Athenian democracy involving the direct and active participation of the masses," and that James "lauded tragedy for its mass affiliation."[96] James argued that the masses in attendance at the Greek tragic theatre had powers of decision and aesthetic judgment. Calling the Athenian audience the "masses" is obviously a prochronism, and James's characterization of what the Athenian audience did as a "direct and active participation of the masses" omits what we know to have been the exclusion of slaves, women, and others from that democratic "mass." Nevertheless, Glick says that for James the chorus becomes "the organizational structure appropriate to actualize mass-driven systemic change," and hence the tragic genre is a medium to "narrate black radical movement."[97]

But it is also interesting to look at the few pages that C. L. R. James added for his 1962 revision to his 1938 opus *The Black Jacobins*. This passage is heavily relied upon in scholarship to prove that James indeed meant the binding of the tragic form to the Haitian revolution and to Toussaint. But I read the passage as something of the opposite. These few pages interrupt his narrative of the Haitian revolution after a detailed and intimate description of Toussaint's final, disastrous mistakes and the moment in which he sees the arrival of the French expedition launched by Bonaparte to restore slavery in Saint Domingue. In the pages inserted for the second edition, James describes the tragic stature of figures such as Hamlet, Lear, and Ahab—tragic because they "assert what may be the permanent impulses of the human condition, and their defiance propels them to heights which make of their defeat a sacrifice which adds to our conceptions of human grandeur." According to James, however, Toussaint's life is "not truly tragic."[98] In this claim, he therefore disputes Toussaint's defeat and imprisonment as "universally looked on as tragedy."[99] He cites the vacillation, indecision, and failure to retain the support of his base in the black ex-slaves of Haiti that overtook Toussaint, and seems to warn against assigning to this a "tragic flaw" as "we have constructed from Aristotle," such as moral weakness. He cannot be retrieved for the classic Greek tragedy. And here James moves from the imaginative universe of fictional tragic heroes to specify that Toussaint's "flaw" was "a total miscalculation of constituent events." That is, Toussaint's defeat was caused by a historical reality, "the historical actuality of his dilemma."[100]

Here James seems to definitively separate the classic genre of the tragic from the historical world, and to specifically resist the designation of tragedy to that world. Toussaint's story does not fit the niche of trans-historical tragedy wherein "the Greek tragedians could always go to their gods for a dramatic embodiment of fate, the *dike*, which rules over a world neither they nor we ever made." James adds to his definition of tragedy a sense of the lack of human initiation, of passivity in the face of the foretold outcome. By contrast, Toussaint had no such resort. Bonaparte, bent on the re-enslavement of the blacks of Saint Domingue, James says, was Toussaint's "fate." James says that "not Shakespeare himself could have found such a dramatic embodiment of fate as Toussaint struggled against, Bonaparte himself." It is clear that Bonaparte, as an obscurantist political reality complexly interwoven into the revolution by slaves that freed the slaves, is not that pre-determined obstacle against which the hero was bound to have to struggled. Bonaparte's entry into Haiti is a consequence of Toussaint's errors at political points of decision. James's inserted paragraphs end in the contra-tragic assertion that no imagination of the tragic genre could have accounted for the historical entrance of ex-slaves, as a chorus, "as the arbiters of their own fate."[101] That is, the tragic genre does not take into account the initiation by an oppressed people of their own freedom. James thus shifts the terms of a genre canonized by European culture to *history*, thereby, as Lisa Lowe says of both W. E. B. Du Bois and James, contributing "philosophies of history rethought in terms of black struggles for emancipation."[102]

The Tragedy of King Christophe

As we move to Aimé Césaire and a discussion of his play, we move to other commentaries on James and on Toussaint. Gary Wilder, for instance, draws from several passages scattered in the latter part of *The Black Jacobins*, including in the paragraphs added for the revised edition. He presents James as ambivalent about Toussaint, ultimately judging him in harsh terms, in order to contrast James's evaluation with that of Césaire. Wilder cites James's statement that the "revolution passed Toussaint by"[103] and that Toussaint's "strategy [was] trapped in the past." Wilder suggests instead that it "anticipated the future" and was "a canny reading of the political present." Following Césaire's vision of Toussaint in Césaire's highly influential *Toussaint Louverture: La Révolution française et le problèm colonial*, Wilder writes that "Toussaint pursued arrangements that would ensure the greatest liberty for his people given the existing political landscape."[104] The politics he sought "would have fundamentally reconfigured the colonial character of Saint-Domingue, the imperial relation between France and the colony, the republican character of the nation-state, and the national character of the republic [and] undermined the racist norms governing the existing capitalist, imperial and interstate systems."[105] Wilder suggests that Toussaint's ideas of an "unprecedented federal partnership" would have been possible only in a world in which these partnerships had already been instituted. At the same time, this paradoxical situation was not as paradoxical as all that. He *had* made a new society in Saint Domingue. The conditions at this site of the new present *were* in place so that "Louverture's regime was an actually existing impossibility."[106]

Aimé Césaire was born in 1913, a decade after James, in Martinique. He moved to Paris to go to school at the Lycée Louis-le Grand. He was a poet and a playwright, perhaps best known in theatre circles for his adaptation of *The Tempest*. He was also a public intellectual and a politician, a communist representative from Martinique to the French National Assembly. Nesbitt, against the general belief that Césaire eventually disavowed communism, argues for his "career-long fidelity to a nonaligned or generic form of Communism."[107]

The powerful idea of negritude, which was rooted in the theory of Martiniquan writer and philosopher Jane Nardal, was developed by Césaire along with his close friend and colleague Leopold Senghor.

Wilder describes the situation of the development of negritude as follows:

> During the late 1930s while studying in Paris, Césaire participated in overlapping social circles drawn from expatriate colonial students, a transnational black public sphere, and the literary avant-garde. These were the spaces from and in relation to which he helped to fashion the cultural and political project that would become known as Negritude. The latter aimed to refigure imperial France as a plural society within which cultural particularity and political universality could be reconciled.[108]

Wilder says that Césaire and Senghor believed that imperialism had created the means for transcontinental cooperation and collaboration. Like James (and Toussaint), Césaire thought of Antillean revolution as a force that could transform France, saying "I know of only one single France. That of the Revolution. That of Toussaint Louverture. Too bad for the Gothic Cathedral."[109]

After moving back to Martinique in 1939, where he taught at the Lycée Schoelcher, he, his wife Suzanne, and Rene Menil founded *Tropiques*, an exceptionally important interdisciplinary journal that conjugated Antillean specificity with European modernism. Like his *Notebook*, this journal was less interested in criticizing modern universality from the standpoint of black particularity than in refiguring the relationship between universality and particularity by fashioning an original Antillean modernism that was simultaneously rooted and cosmopolitan.[110]

Césaire traveled to Haiti in 1944, where he became further convinced that national sovereignty was not a political solution to a postcolonial situation. It is possible to say that the politics enacted by Césaire are in a sense a resurgence of the new present of the Haitian revolution, in that his vision of cooperative federalism was influenced by Toussaint's constitution. For Nesbitt, Césaire was the exemplary Black Jacobin, who "like Louverture before him . . . remained faithful to the French state in its egalitarian, Jacobin democratic tradition."[111] Aimé Césaire's first work on the Haitian revolution was *And the Dogs Were Silent*, a historical study, but censorship in Martinique prevented its being produced there. It was eventually published in 1946, with Toussaint and the Haitian revolution per se excised. In 1953 Janheinz Jahn, a German director and ethnographer of Africa, worked with Césaire for three years to ready *And the Dogs Were Silent* for the stage, finally producing a version in 1960. It was interpreted at that time to be a reference to the Republic of Congo and Patrice Lumumba.

The Tragedy of King Christophe is widely considered a masterpiece in Europe and the Caribbean and is frequently performed. It was first performed at the 1964 Salzburg Festival, directed by Jean Marie Serreau. Translators and editors Paul Breslin and Rachel Nye say in their 2015 edition that its lack of production in England and the United States "owes much to its resistance to translation. It encompasses an extraordinarily wide range of stylistic registers, ranging from lines adapted from Racine and Lamartine to songs in a Martiniquan-inflected Haitian Kreyol; from the derivative neoclassical verse of the court poet Chantallete to the surrealist poetic soliloquies of Christophe."[112] They go on to describe its multiplicity of behavior forms and character types. Their offering in this edition is vital, therefore, as it discloses what is a beautiful, strange, and strong play, one that unlike *Toussaint* is clearly the work of a gifted and skilled playwright.

The editors imagine that although Christophe committed terrible crimes he is entitled to be a tragic hero because of the almost impossibly complex difficulties of the colonial situation in which he found himself. They also cite David Scott, relying on his concept of "emplotment." In their reading of Scott, "when the future arrives, it is not necessarily one projected by revolutionary hope, and new forms of emplotment are necessary to absorb the unexpected turn of events into a compelling narrative."[113] Presumably, they see this as the tragic emplotment that they think characterizes *Christophe*. They rely here on two closely linked temporal assumptions, narrative and plot, both of which must be refashioned to tell the story of revolutionary degeneration in the absence of revolutionary hope. Emplotment here evokes its close phonetic relative, entrapment, by means of which what has happened, a history, a history of initiation, is re-scripted into a generic treatment of historical revolutionary failure as a compelling story.

In contrast to Breslin and Ney, Wilder uses Scott's work to underscore his own interest in the temporal conditions of emancipatory possibilities. Scott, in *Conscripts of Modernity*—the book that preceded *Omens of Adversity*, which I have already cited—presents the argument that the postcolonial imaginaries of the nation, as that which follows revolution, have failed and that thought should no longer be directed toward that now-superseded historical project. Instead, all the questions directed to the past in expectation of answers should be reexamined, because those answers may no longer have any value for the present. Wilder says that he agrees with being attentive to the politics of our contemporary situation. But Wilder differs from Scott in his refusal to be taken up with "futures whose promise faded after imperfect implementation" and to turn instead to "alternatives that might have been and whose unrealized emancipatory potential may now be recognized and reawakened as durable and vital legacies."[114] In particular, Wilder says, Césaire and Senghor "were especially attentive to the complex relationship between politics and time. They explored separately how inherited legacies may animate current initiatives and how present acts may liberate the not yet realized potential sedimented within reified objects."[115] Rather than considering the play we are about to study as a tragedy, a compelling emplotment of revolutionary failure, we can take it on with Badiou, and with Wilder, so that as with James's play our outcome is an affirmative gesture.

The play is indeed a marvel of languages and references. It uses farce, parody, Haitian folk characters, Shakespearean style fools, heroic speech, song and music, and colliding and historically inaccurate time signatures to tell a sort of narrative tale beginning when Christophe, leader of the north of Haiti, finally breaks from the mulatto Pétion, in charge of the "republic" of Haiti in the south. Christophe, a slave with a wide-ranging history, including fighting in the American Revolution and returning to Haiti to become a cook, ultimately deserted Toussaint. After Dessalines, who took over after Toussaint was imprisoned and called himself Emperor, was assassinated, Christophe took power. Christophe, refusing Pétion's offers of an empty figurehead position, characterizes himself from Pétion's point of view in a vivid time image, as a mechanical clock. This marks from the very beginning Césaire's attention to temporality. Christophe says, "Yes, yes, my masters, I know that, under your Constitution, Christophe would be only a cheerful fellow carved in ebony, a jolly jacquemart, busily striking the hours of your law on the clock of his impotence with a ridiculous sword, for the crowd's amusement."[116] There is a sneer in Christophe's address. There is the contempt, the tension that remains in this postcolonial, post-revolutionary nation seeking its political form. We can hear his sense of his time as something that will be partially outside the binding of mechanically progressing time, unquestioningly marked as it passes.

In 1811, Christophe was crowned King of Haiti, changing his name from Henri to Henry I, evoking the power of the English dynasty. The next years were characterized by battle after battle with Pétion's forces while a rebel force gathered, disgusted with both Pétion and Christophe. In the play, Christophe is characterized from the beginning as imposing censorship of speech, indulging in grand self-presentation to a laboring class of peasants (those who were once slaves, from whom he demands absolute loyalty), obsessed with the need to prove that the new black nation and its black people is the equal of any European "civilization," and obsessed with time. His pomp and splendor is from the beginning set into tension with this peasantry, who are given by Césaire a kind of constantly mumbling, insubordinate tone, a sense of a perception, and a history, from which Christophe has, from the beginning, withdrawn. Césaire gives them the title of citizens, a perhaps ironic marker for their short-lived inclusion in the Fraternity of the French Jacobin revolution that declared the slaves free, or perhaps a marker for their actual political acuity, their political capacity. For instance, early in the play, "Second Citizen," commenting on Pétion, says, "They say that to get the French king to recognize him, he's offering to pay reparations to the former colonists! A Black man offers to pay reparations to those whom Blacks so rashly deprived of the privilege of owning Blacks!"[117]

Christophe addresses the "citizens," the crowd assembled through Christophe's agents, at the end of Scene 2, introducing the theme that will be at the core of the play: work versus indolence, the abolition of the state of being tired, of exhaustion. He speaks of their "hatred of discipline" and of "sensual torpor." He says that after ten years of black freedom, none of this can be allowed to squander the "treasure that our martyred people amassed in a hundred years of labor and blows of the whip. Which is to say that with me, from now on, you have no right to get tired! Go sirs! Go, disperse yourselves!"[118]

Hugonin, the Fool character, is laced through the play like a strategic poison, commenting, playing, messing, twisting—keeping the play on a kind of toxic edge, an edge that prevents Christophe from the completely successful self-presentation of his absolute authority. He is instrumental in the next scene, which is parodic, bizarre, and ridiculous. A "Master of Ceremonies" has been called in to "rehearse" the coronation ceremony. According to Césaire's footnote, this is "a white man sent by the TESCO (Technical, Educational, and Scientific Cooperation Organization) in order to provide technical assistance to underdeveloped regions."[119] In this barely concealed reference to the despised developmental agencies with which our world has become increasingly saturated, Césaire frames the content of the scene as an imperialist intervention. The "courtiers" mock their new aristocratic designations—Duke of Marmalade, Duke of Lemonade, Duke of Candy-Ditch—and perform "apelike, ironic contortions" as they spoof the comportment lessons given by the MOC.[120]

The lessons are meant to substitute the white aristocratic body in the place of their own black ones. Hugonin slyly comments to the also highly skeptical General Magny (now Duke of Plaisance) that

> It was inspired, do you understand, inspired, this idea to invent a nobility. For the king, it's a way to baptize whomever he wants, and a way to become every Haitian's godfather. Of course, if the husbands let him, he'd be everyone's *father* instead . . . . Now then, to try on my role for the first time, I propose that the child the king has made with a certain plump lady you know of should be called the Duke of Varieties.[121]

Here Hugonin establishes another theme of the play, which Césaire develops throughout—that of Christophe's masculinist, patriarchal vision of "equality." Christophe "pats several rumps" as the ladies enter, calling them "Madame Syringe, Madame Little-Hole, Madame Knocks-Out-Your-Eye."[122] He is insulting to all, abrasive, belligerent, over-seeing with a cruel humor. He practicses reaction to the revolution. Anointing himself as the father who will manage everything for his children is a symptom of this. Clearly, no form of equality can exist in this relation.

And yet, as is typical in the play, buffoonery, cruelty, smallness, limited vision, and ego are suddenly interrupted when Christophe begins to think, to imagine, somewhere within the very megalomaniac vision that sustains his excessive energy, the political situation he is after, one that in some ways, in glimpses, seems true to the idea of the revolution. His megalomaniacal vision is simultaneously one of re-birth and re-naming. He speaks of "thousands of half-naked blacks / the sea vomited up one evening," blacks "shaking with a power to speak / To do, to construct / to build / To be, to name, to bind, to remake."[123] His vision is absolutely for the equality of black people within the world context. His hallucination is of the country (under his paternal care) rising from the ashes, with the phoenix as the symbol on his coat of arms, as well as becoming the symbol of Haiti. Upon being crowned, Christophe reiterates the foundational core of Toussaint's leadership, saying that he will "never permit on any pretext whatsoever the return of slavery or any measure contrary to the freedom and full exercise of civil and political rights by the people of Haiti."[124]

Césaire points here to the making of a world from a post-revolutionary situation. This is an action of creating that does not necessarily belong (only) in the present of the vexed outcome of a revolution thwarted from many sides. Here, decisions are still being made in the experiment of discovering a new political situation for Haiti. If we pay attention to time, as Césaire is doing, we are attentive to inner rhythms of the traces of revolutionary time, after the event of the revolution itself, and what forms they might take. We do this instead of making an analysis based on historical forms of narrative, like modernization, or aesthetic forms of narration, like tragedy. Time itself becomes existent in a new way. There is a time in the play, all through it—the play is saturated with it—which belongs to the revolution.

And yet, Christophe's terrible conflict is with time. He feels time acutely. "Time has us by the throat!"[125] What he feels is his own frantic worried dissonance with that revolutionary time. He mishears it. He finds it hard to be true to revolutionary time—he is panicked by his own sense of a time that is happening without him, without Haiti, beyond him. He cannot listen for what the time is that will allow Haiti from here on out to continue to explore the consequences of the new present, the revolution. He has become temporally distorted. There is a time, specific to the revolution, that he cannot hear. It is not chronological, modernist. This is in the play when Hugonin says to Christophe: "The people live from day to day, your Majesty." Christophe replies: "I wanted to solve the enigma of this people that always drags itself down!" Hugonin responds: "The people go at their own pace, Your Majesty, their secret pace."[126] Christophe has missed the revolutionary pace of his people, the pace of these slaves who are slaves no more. "The people" as a body that has formed in the new present generates its own time.

As I mentioned in Chapter 1, Kristin Ross writes that during the days of the Paris Commune, the shared form of address, "Citoyen"—itself born in 1789 and "kept alive thanks to secret societies and revolutionary traditions"—was in its differentiation from the bourgeoisie mesdames and messieurs, an address to people who had separated themselves from the "nation" and therefore created a "gap or division in the *now*." Ross writes that with the appellation they create a new temporality in the present, different from the time of the nation, one that belongs to a "now of a shared political subjectivization."[127] In fact, the temporality Christophe disparages in the citizens/peasant/laborers/ex-slaves may be an inseparable part of their revolutionary capacity. Their ongoing emancipation could not be gained through hurried up forced labor so that the Haitian nation could generate institutions identical to those of the Europeans, including ones that would then teach the

Haitian illiterates to be good citizens. Instead, as Ross describes the Communards, their temporality was in the way that "their daily workings inverted entrenched hierarchies and divisions."[128] The play is saturated with the drumming in the forest contrasted constantly to the militaristic drumming of the war, with Christophe's inflated arrivals and entrances. It is the sound of the paces of those who are still members of the revolutionary body, in their everyday. Voodoo drumming, in its polyrhythmic complexity, is the sound of revolutionary time, pace, the people's secret pace. Far from the drums sounding as the threat from "primitives" living in an eternally fixed world outside modernity—as in the forest drumming in Dudley Nicholas's film *Emperor Jones*, for instance—these are people who have initiated their own history in their own revolutionary temporality.

Christophe, in a frenzy near the end of Act II, is faced with a spokesperson for the peasants, who comes to tell him that the people are tired, exhausted, that they wish for rest as "the sweet hope that all classes of our society will finally enjoy." He roars at them that he "would hate my victory at once if it encouraged you to be lax" and that he will "give them no quittance."[129] He compares their experience to going through a raque, "the enormous crevasse, the endless trough through the mud." Referring to a specific raque along the Artibonite river that he wants to turn into the Nile of Haiti, he says it is always "packed thick with mud" and "this century is rain, a long march through endless rain. Yes, in the *raque*—we are in the *raque* of history. For black people getting through it means freedom . . . . All right, then, do you get me: you don't have the right to be tired."[130] Calling the "delegations" back, he suddenly outfits them with farm tools over their shoulders like weapons and orders them to march, an image of the militarized labor that Toussaint started and that Christophe enforced in the extreme. For Christophe, history produces long, slow, miserable slogs for the oppressed, and their only destination can be at the end of this slog, stepped to a militarized beat. Only after the long, long course can they be free. Again, Christophe does not hear the temporality of the peasants, rebels remaining loyal to the traces of the new present, their experience of revolutionary time.

In a brilliant turn-around, a departing peasant mumbles to Christophe: "Funny idea, to go walk in the *raque*. You walk alongside the *raque*. Everyone knows that. The *raque* is a trap. You better off taking the river—leave all the muck to the left and right, cut through by the river. Nuncle, you'd have to know the rivers."[131] The peasant here refers to a peasant (revolutionary) capacity (you would have to know the rivers) linked to a form of temporality distinct from Christophe's modernist, developmental time. The peasant knows you do not do time like that: that is not the revolution.

I referred at the beginning of the chapter to the way in which John Patrick Walsh turns from Césaire's analysis of Toussaint to evoking the word "tragedy" ceaselessly throughout his work on the play. Walsh's way of describing the peasants in the play, for example, and in stark contrast to my own, refers to "the renaming of the downtrodden" as one of the "timeless problems" that present themselves in Christophe's Haiti, and joins the peasant to the trope of nature in the play, citing the peasant's "wisdom and patience to respect her."[132] He refers to the "voice of the folk" and their "popular wisdom,"[133] and how they like to "take time off to enjoy the land."[134] Rather than paying attention to the fact that "the peasants" are made up of either revolted slaves or their children, a people who successfully fought for their own emancipation and therefore carry the embers of that revolution in secret, in hiding, Walsh says of the scenes with the peasants that "the dramatized suffering of the peasants *mediates* another way to read their plight (and periodic revolt) under Toussaint."[135] Walsh is here referring to militarized labor. For Walsh, their situation is tragic. For Walsh, "plight"—that excessively used liberal humanitarian term to designate to world's poor as sufferers, rather than political beings—is the useful descriptive term.

But the peasants in the play are actually canny, strategic people, facing the effects of Christophe's ambivalent and ambiguous moves between fidelity, reaction, and obscurantism in relation to the revolution. Still in some ways a faithful subject, Christophe also attempts to institute an extinguished present, a measured present. For Christophe, the catastrophe potentially contained in the consequences of the new present is the tendency of the freed slaves to indolence, sensual stupor, and taking their rest. In another peasant scene (Act II, Scene 1), two peasants having an allotted rest are talking about the "time" (meaning the political climate), discussing in a sort of code the way that they have been excluded from the land that they freed from the plantation owners, the whites. One concludes that they must trust in God, and the other responds that they cannot get a prayer up to him because they are prevented by the Royal Dahomeyans from the drumming, which is a form of worship. The Royal Dahomeyans arrive, positioning themselves to make an announcement. The stage directions indicate that the voice of the reader of the proclamation from Christophe is underscored with the sounds of drums, which grow to the point of "obsession." The Dahomeyan declares that from now on agricultural work will be disciplined and punished, just as if the peasants were in the army. They will fulfill their duties with "precision, submission and obedience, as soldiers do." Finally, the Dahomeyan reads, "liberty cannot exist without labor."[136] It is an act near to that of full obscuration. Christophe, through the Dahomeyans, seems to draw the idea of equality and liberty under the bar of a discipline that will render those principles once again inexistent.

In Act II, Scene 2, a group of "ladies" are talking in a drawing room. One is telling the story of a man who was caught sleeping by Christophe at an hour when, according to the Code Henry, sleeping was not allowed. The scene moves into a flashback of the scene in which Henry, seeing the man sleeping through his field glasses on the high wall of the Citadel, has him shot dead with artillery fire. Christophe rages against his aide's suggestion that the man is exhausted. The man must be obliterated. As the scene continues, the ladies mark the way in which Christophe has dragged the revolution under the bar. One says to Vastey, who has joined them, that the situation "looks horribly like the same old thing the you [Vastey] once fought against." Vastey equivocates, saying: "Eh. Well, sometimes history can go forward by only one road. And all must take it." Another lady retorts sarcastically: "So the path of liberty and the path of slavery would be one and the same." And the original lady follows up with, "A charming paradox! In effect, King Christophe would serve liberty by means of servitude."[137]

And yet, even given these crushing ironies, Césaire is marking the new present. As Wilder says, Césaire wrote the play to engage with "the politics of freedom and the problem of time."[138] Christophe is a reactive force, but to make him a straightforward tyrant would be to simplify temporality—instead, as I have suggested, it is as if he is battling the temporal world of the revolution in his own self, struggling to make sense of his position in relation to time. His is a character filled with the dynamism of new time, even as he converts it into a regime of bodily labor that is machinic and brutal. Time is converted to the measure of productivity in brutally laboring bodies, even as Christophe feels, senses, that this is not the time of his own revolutionary conversion, the new present. Césaire is directly confronting the problem of time as time in a post-revolutionary, postcolonial situation. What is the time that we should hear, be faithful to, in any new present?

Césaire provides another graphic example of Christophe's confrontation with time in Act II, Scene 3. Christophe, building his Citadel, furiously increasing the pace of construction, orders that women and children will now have to help build it, a terrible labor of carrying stones up to the top of the promontory. For Christophe, the temporality of modernity is a razor challenge to his stuporous laborers. He can only feel time slipping away, the fingers of time around his throat in this world he imagines as full of his people working toward a common goal: to monumentalize, to make a monument for, the equality of black people in the world, in a world in which they have lost time, have to make up time. Despite his own brutalities, he condemns a man who was caught whipping his peasant to be publicly cut apart. Christophe says, "I put workers at his disposal. I didn't give him slaves," and that "it's time to talk sense into those blacks who think that revolution means taking the place of the whites, yet continuing in the same way—that is, on the backs of us blacks—to play the white man."[139] He either fails to see, or deliberately overlooks, how closely the temporality of the modernity with which he is competing drives the laborer back to slavery.

Although Césaire does not in any way overtly situate this play within the development of capitalism through the plantation economy in the Antilles and the United States, the modernity into which Christophe hopes to slip as a challenge to the white world is a modernity being shaped foundationally by developing capitalism. In a sense he drags the Haitian revolution under the bar into its emerging opposite, that economic system that will have lack of equality and the exploitation of labor at its core. This includes the systemic inequality of women. In Act II, Scene 4, Césaire writes what can only be described as a slave market. Christophe, convinced that all the people are "libertines," decides to perform enforced marriages to stabilize the family structure and to stabilize especially women. He puts Hugonin in charge, and Hugonin deliberately does the work with the language of the slave market, including the possibilities for reproduction of more slaves, of "breeding." Looking the peasant men and women over, he says:

> Let's see—will she do, that one there? Yes, wouldn't she? A little bit fat! But the fat ones are the best ones! She's yours . . . . Sold! . . . . There's someone for every taste . . . . You, you've got a build like Hercules . . . . And that woman there looks like she's got a strong back . . . . Take her . . . . Come, ladies and gentlemen, agriculture needs hard working arms and the state needs soldiers . . . . Well then, I'll put it to you like this: good night and plow hard![140]

There is a clear sense of closing up the space that we could say is opened by the new present, condemning it. With his Citadel, Christophe in part is replacing the new present not only with a time in competition with modernity's time, but also with monumental time. He is stabilizing time along with the family, the women. He has a feverishly driven attachment to monumental time, a time that drives people literally to death in constructing the monument. He is creating fixed time, time in stone, time presented as eternally unchanging. The image of the monument also has the echo of incarceration, the tomb. Christophe has Brelle (the Archbishop who has come to him asking for retirement back to France) killed and then walled up in his palace, which he refers to as a tomb.

In the scene directly following this, we see the horrifying labor of constructing the Citadel directly. It is a scene of a storm, wind, slippery stones, utter exhaustion, and the overseer, who may name the situation of incarcerated chain gangs or slaves. The laborers sing that they are running away, going fugitive, back to their huts, when Christophe enters. The overseer pleads with him to send the laborers home, pleading that the stones are slippery, the laborers are slipping backward as they try to hoist up. Christophe refuses and instead tells him to get fifty more workers out into the storm. Christophe himself picks up a trowel and says, "Look, I'm going to show you how a nigger of consequence works!" The laborers are now singing their work song, exhausted: "From now on we refuse to die / From now on we won't die."[141] The storm grows worse and there is a tremendous explosion even as Christophe is in a hallucinatory grandiose state in which he now appears to mix the voodoo drum with the military one, as if he were Shango, the syncretic god of lightning and war. He interprets the storm as if it were an attack and calls for the overseer to make the drums sound to make the storm even worse. He whips himself into a frenzy, even as he is informed that lightning has struck the treasure house and its stores of gunpowder, destroying a whole garrison. It is a gruesome and almost hallucinogenic scene.

Act II, Scene 7 is an exquisitely complex scene, as I have already mentioned, and to which I wish to give more detailed attention. Again, the scene is the banquet after his coronation, which, according to the stage directions, takes place in Christophe's villa. The scene is set up by the Master of Ceremonies (TESCO). It is a bizarre farce, full of the tension with the European signs of royalty, including a casket with royal eating implements, the priest, and the court poet, who is challenged to praise Christophe by the priest, Archbishop Corneille Brelle. They take turns, competing over the languages with which they fawn, Latin versus European heroic verse, goaded by Christophe to do so while it all becomes a European poetic babble, a parody. Hugonin (as Shakespearean fool) plays dog. Christophe declares that he will not engage in "servile imitation" of European forms. Instead, "if we are to raise this people to civilization . . . we must also let the national spirit speak." He is at pains to point out that they are out on the veranda, not in a room in the Tau Palace, as the Master of Ceremonies has suggested they imitate, drinking rum, not champagne, "Haitian style."[142] Chanlatte, the poet, suggests that rum be the national drink and makes up a song about it, whereas Christophe declares it will be the national drink. Hugonin objects that although he is "as patriotic as anyone else. Anti-white as anyone else," he likes champagne. Christophe says "Champagne, champagne, you worthless guzzler!" and throws him a bit of food to "help you forget about champagne."[143]

There is a national self-fashioning going on here, but in such a parodic key that it ridicules the idea of the nation. It toys with the defining differentiations between European forms and this new postcolonial nation, even as Christophe is literally embodying a European monarch (Henry I) and demanding the same semiotics that inhere in a European court. He himself seems aware of this irony and to be playing on it. Hugonin helps. After the Archbishop Brelle and the poet have delivered their peons to Christophe, which he mocks, Hugonin bites his leg and parodies the fawning attention of a dog. The scene to this point is a tense mess, a knot of colonial mimicry, a density of the problematic of what we would call today humanitarian intervention and intervention by World Bank in a destroyed country. There are decisions about what to take and what to reject, a leader who seems to sense an insurgent black modernity to counter European modernity, a sense of an exceptionality of black experience, the tension that he embodies between the political form of a nation (republic) and that of a monarchy, the tension of the form of monarchy as a European institution and as an African political form.

The scene changes in tone on the hinge of a message brought to Christophe, a letter from his friend Wilberforce in London. William Wilberforce was an English philanthropist, a leading abolitionist, and also an evangelical Christian, who worked with the famous abolitionist Granville Sharpe. It is a letter congratulating Christophe and telling him that Wilberforce has enrolled him in scientific and Bible societies in England. This earns a sharp laugh from Archbishop Brelle, to which Christophe retorts sharply that there is nothing wrong with that. But here the scene moves into its second and quite different movement, as if Christophe is done with parody, mockery, dressing up, trying on, playing in the knot, and is triggered back to his central theme.

Christophe rejects the letter, not on the basis of his induction into English learned societies by his abolitionist and philanthropic "friend," but on the advice that Wilberforce and those like him give. It is an agricultural metaphor for the "gradual ripening, year by year" of the nation. It is advice about time. It is advice about time that resorts to a processional definition of history, in the guise of natural temporal unfoldings. Christophe rejects it because it is not his sense of the temporality of nation building. He begins by quoting the advice: "'We must give time due time.' But we don't have time to wait when it's precisely time that has us by the throat! To entrust the fate of a people to sun, rain, and the seasons, what a strange idea."[144] He rejects the nature-based metaphor that most likely is founded on a colonial idea of the former slaves as closer to nature, and moves his time more centrally to modernity's time. At the same time, he names his own time against the liberal abolitionist condescension.

This is also one of the only scenes with a significant inclusion of the only female character in the play, Madame Christophe, who is present at the party. Interestingly, the scene narrows down to an intellectual battle between her and Christophe, conducted in sophisticated terms by both, even as she becomes one more obstacle to his vision. Adopting a strategy of self-diminishment that is actually aimed at Christophe's inflated view of himself, she minces no words in cautioning him, as a "poor woman," a "simple woman," who somehow finds herself with a crown on her head. She warns him against "spurring the fiery horse" of his heart, a warning with specific temporal implications, and predicts ruin ahead, especially for her children, because of their father's "excess." In reaction to her, Christophe explodes into an important and complex speech that insists on the foundational difference between white and black people in the face of philanthropists, as "irritating," he says, as "the talk of pro-slavery hacks" who say that "men are men and there's no such thing as whites and blacks." He signals the way in which one of the signal slogans of European enlightenment and modernity, the doctrine of equal human rights, obscures the black and white Jacobin axiom of liberty and equality. He almost viciously attacks, saying, "Who will have us believe that all men, I say all, without privilege, without special exemption, have known deportation, trafficking, slavery, collective debasement to the status of beasts, total outrage, enormous insult, which all who have suffered it wear, plastered over their bodies, their faces: the all-annihilating spit! We alone, Madame, you understand, we alone, we blacks!"[145]

But even as he excepts black colonial and slave experience from the Enlightenment's self-soothing banalities, he takes the bait. For him, this exceptionality becomes the command for the exceptional amount of work and effort on the part of those same blacks to climb out of the ditch into which they have been thrown. He says, in words that have a familiar echo from the history of race in the United States, that black people have to work harder than everyone else. He wishes, through the brutal, militarized labor time, to institute black equality in the image of white modernity, through what he hopes will be the exceptional capacity of the black people of Haiti to work harder than anyone else to prove that they are equal to the white people of Europe's "civilizations." What is afoot in the play, sensed, what he senses, is the potential for a black modernity, with a different temporality, marking it as a new difference in a history initiated by a slave revolution. But the lure of proving equality on terms established by white Europeans, in terms of European education, architecture, and culture, is unbearably strong to Christophe.

The scene continues with a persistent Madame Christophe continuing her strategic aimed self-diminishment, in effect allying herself with the peasants, the workers, the black "uncivilized." She says, "Christophe, do you know how, in my little / Woolly head, I think of a king?" She contrasts the Mombin tree under which creatures can take rest to "the great fig tree" that "strangles" all the plants growing around it. Christophe corrects her by saying that the tree to which she is referring is the fig tree from the New Testament in which Jesus curses a fig tree for not having fruit for him when he was hungry and the fig tree withers. It is used as a testament to the power of faith. It appears that Christophe leaps on this opportunity to correct his wife's analogy with the proper parable. He appears to frame her comment as an ignorant mis-citation, another clear instance of why he must work the blacks so hard toward their education and equality with Europeans. He says, after correcting her, "think about it dear wife! Ah, I demand too much of blacks?"[146]

Confronted with his wife's reaction, dissatisfied with the simpering flattery that he himself, in a parodic mood, solicited, Christophe erupts with agitation at hearing the beat of the tam-tam. He feels threats, traps, an undermining and weakening of his authority. In one fell swoop, he dismisses women, priests, and courtiers. One person stays behind, Martial Besse, Christophe's French engineer. Christophe sneers at him for having no ideas about how to reassert his—Christophe's—power. Besse, under pressure, arrives at the suggestion of patrimony, which he thinks will be a kind of ideological means for Christophe to drive his people harder, "awakening in them their hidden power."[147] Christophe will be creating for his people their patrimony, their inheritance from a father. This idea seizes Christophe's imagination immediately. His tangle in his gendered, patriarchal institution of his power, his idea of self, of what designates and secures his leadership, especially having just had a tantrum in response to his wife's measured and intelligent criticisms, snares him to another temporality that is antithetical to the new present, that of a succession of hand-me-downs through time. What he is to do is not true to the traces of the new present because it disavows the potentially ongoing creative time of that revolutionary trace with the enforcement of the father's power in relation to time: to construct history through his wealth and therefore to subject temporality to the power of a single agent. It is the father who will awaken the people to their hidden power, rather than the people who, becoming existent, explore their own newly usable capacities.

Christophe thanks Besse profusely for his suggestion. He seems to have a vision. He speaks of the "swelling earth" as "the first step out of chaos, the first stride to the sky." In a hallucinogenic distortion, Christophe begins to see the Citadel he will build on the very mountain on which he and Besse stand. It will be "one hundred-thirty feet high, twenty feet thick in the walls, of limestone and the ashes of cane trash, limestone and bull's blood." It is not to be a "palace" or "a strong house to protect my goods." It is to be a Citadel, by which he means "the liberty of all the people. Built by the whole people, men and women, children and elders, built for the whole people." It will be impregnable, a "massive breastplate of stone."[148]

In this scene, the coronation scene, and in this last speech in particular, Césaire ignores the factual chronology of Haitian events. The actual Christophe commissioned the Citadel in 1805, and its construction was not finished until 1820, the year Christophe committed suicide. The coronation took place in 1811. Although he is still in his Haitian-style house for the coronation scene, his stupendous palace, San Souci, modeled on a Renaissance villa, was under construction, started in 1810 and finished in 1813. Césaire thus collapses the times of the historical document into a crash of languages, forms, tempos, rhythms, allusions. Inside this thicket, this collision, there is Christophe.

In reality, as in any situation, it is not only through his agency that the new present has been obscured, drawn under the bar. Perhaps the first event in this direction was the decision by the body, as articulated by Dessalines, to found the nation, that which the internationalist Toussaint (in James's hands), as a member of what we might call the Jacobin body, refused to do. This is what it feels like, looks like, when the new present is obscured. This is what you can hear around the edges, inside, from the exterminated, when the new present is obscured. The symptom of the obscuring of the new present is emotional and temporal disfiguration, contortion and pain. As I have said, Christophe in his own way fights against this obscuration and remains absolutely committed to the egalitarian axiom on which the new present in Saint Domingue appeared: the slaves can be and are free. But that new present in which that axiom, that truth, was active is (temporarily) consigned again to inexistence. Slavery turns to militarized labor, something that happened under Toussaint. This temporality, this new present, is both a thing of great force, terrifying plantation culture across the Caribbean and in the new United States, and a thing of delicacy and situational impermanence. This is what the play shows, along with the deeply affective and emotional effects of the process of its disappearance. It shows the torment, the obsession, of living in "the extinguished present."[149]

But the play also shows the retention of the traces of the new present. Again following Badiou, we could say that there are signs in the play that "a fragment of truth inserted under the bar by the machinery of the obscure can be extracted from it at any instant."[150] For instance, early in the play Metellus, the head of the people rebelling against Christophe, has been captured and brought before Magny. Metellus, in a beautiful speech, recounts his ongoing fidelity to Toussaint and the revolution. It is a tale of hunger, cold, terrible weather, and exhaustion. Metellus describes this army as having "seized fate by the collar, beside Toussaint!"[151] He says:

> We were going to found a country
> That was shared among all of us.
> Not just the landholders list on this island!
> Open to all of the islands,
> To all of the blacks! All the blacks of the world![152]

In this exhausted present, Metellus holds a tiny ember of what has happened.

In this reading, the play is not about (or not only about) the tragic denouement of a tyrannical postcolonial leader caught in a game of catch-up with European modernity. Or, rather, it *is* about that, but there is something else in it too, a different kind of study, a different orientation to revolution. It is about the tension between a new present which came into existence bearing the truth that the slaves can be and are free once again into a newly situated world through the evental uprising of slaves, nourished on the uprisings of other slaves, and the institution of a lusterless present. That is, the new present will always disappear. But the important thing, contra tragedy, is that it will appear again, and that any reactive situation depends upon that very new present that it seeks to suppress. Césaire condenses the situation of this suppression in such a way that all its collisions, its flaming parts and components come into view, its dynamics, its fervor, even its hallucinations.

The scene of the coronation celebration on which I have focused, for instance, in staging its multiple collisions—at the heart of which is the increasing insistence on a subjection to modernist time—imagines the actual event in ways that concentrate on its internal deformations. According to Ada Ferrer, the celebration that preceded the coronation stretched out for days, with festivities in Le Cap such as "dances, musical performances, the public recitation of poems, and invocations of Greek and Roman gods: Apollo, Neptune, Mars, Minerva, Clio, and l'Empire des Fées."[153] Césaire reshapes this bizarre amalgam, imagines its obscurantist quality, but also, in this scene and throughout, features Christophe's fragmentary and inconsistent radical black vision. And although Césaire's dramatic situation is claustrophobic, with little sense of other than a defensive, embattled small world, that vision is huge, stretching to Africa, to the voodoo roots of Haiti, to a kind of world geography. It is a vision that "cancels the slave ship,"[154] as Christophe says, in one of the most beautiful lines of the play. Here Césaire, through Christophe, might be seen as contributing (even through invention) to what Ada Ferrer calls "an intellectual history of the Atlantic World in which enslaved and free people of color are active participants,"[155] a contribution which might itself be seen, contra the allegedly tragic narrative, as a trace of the new present that the Haitian slave revolution created. The historical record, according to Ada Ferrer, studying the connections between Cuba and Haiti, shows that the news of "the coronation of a black king in Haiti spread like wildfire, covered extensively, for example, in the North American Press."[156] In Cuba it was talked about everywhere.

Christophe was in fact a sign of black revolutionary success that circulated throughout the plantation Atlantic, striking fear in the hearts of the ruling classes, and inspiring slave populations. Ferrer writes about the fascinating Cuban Antonio Aponte, a free black, carpenter, artist, leader of a plot for slave insurrection, and contemporary of Christophe, who created a now unlocatable book of drawings and images, "a breathtaking array of stories, allegories, characters: Greek and Roman Gods, European and mostly African kings and emperors, black priests, saints . . . Indian woman, rooster, ships, moons." Ferrer says that there was a particular image of a black figure that Aponte explained to authorities as a "black dignitary in Rome," but to slaves planning a revolt he explained that it was an image of King Christophe. Ferrer concludes that "the book was a . . . subversive experiment in thinking through a black kingdom—the one he and his companions were seeking to create."[157] King Christophe was a key to this project.

In the play, at the end of his life, overtaken by voodoo, poetic, and surreal imaginaries that are concretely pressing upon him, he seems, even in his grandiosity, paranoia, and smallness, to draw near to the revolutionary new present of which he was once a part. Echoing Richard II's desperate, choked, bitter reflection on his heart as clock in a very different key, in an equally great speech, Christophe calls his heart a drum, a "hurricane of blood and life."[158] He imagines himself at the head of the "assembly" fighting again. He says to Vastey, his mulatto general, "just as the earth preserves in its foldings the trace of its past upheavals, you have known . . . no . . . you have *lived* in the red of your hair burning like fire, the infernal breath of the lightning . . . therefore you are black . . . . I anoint you black."[159] Shortly after this, the sound of a shot is heard. He has killed himself. The play ends by calling for that very symbol of what will come again, and again, the phoenix, the red and gold coat of arms for Haiti, Christophe's "immortal coat-of-arms."[160]

I do not mean here to dismiss the critical importance to both playwrights of the question of the failure of revolts, the institutionalization of tyranny by black rulers against black peoples, or questions of postcolonial development. It is to suggest that an emphasis on tragedy and its temporality surrenders the question of revolution to an inevitable destination. Theatre may be a container, a vessel, a medium whose temporal innovativeness can accomplish what the historical narrative and the tragic genre, however intellectually and conceptually well conceived, cannot: a sense of the time that can be in our hands, a delicate vessel for the unfolding of a human capacity that it is in the interests of the victors (in Benjamin's words) to draw under the bar. Theatre may offer a means, in treating history, to avoid being buried in an avalanche of historical ruin, and to instead seek in the past for emancipatory presents and stage them as such. Here they are, again, on our stage, in the living breathing medium by which we have again constituted their promise.

3

Changing Time in the Time Before the End

A free man thinks of nothing less than of death, and his wisdom is a meditation on life, not on death.[1]

To think existence without finitude—that is the liberatory imperative, which extricates existence from the ultimate signifier of its submission, death.[2]

"You know why we're here: we're here to live and die." The omniscient and invisible narrative voice announces this to the child performers of *Before Your Very Eyes*, Gob Squad's performance in which a group of children enact successive stages of their lives and, ultimately, their deaths. The phrase is also among the initial directions given to the children, the reminder of their task. The phrase refers to the scripted actions of a piece of theatre, and to the theatre itself, or what is presumed of it. We are here, in the theatre, to take stock of life lived in its inevitable track toward death, of which the ending of the evening in the theatre is a symbolic and experiential underscoring. As I mentioned in Chapter 1, Matthew Wagner remarks that like a life, theatre is defined by the fact that it ends, that the "identity of any piece of stage work is supplied by its death,"[3] and that the futurity so intrinsic to the theatrical experience hinges on the knowing that there will be an end, a death. By contrast, here I suggest that one of the specific temporal capacities of theatre is to stage what Giorgio Agamben calls penultimate time, the time before the end, before death. I am guided throughout the chapter by what William Kentridge says about his piece *Refuse the Hour*: "We are not going to escape our journey to the black hole at the end, however fast we dance and run away, but that dance and that run are still what it's all about."[4]

In this chapter I will be looking at four theatre pieces that stage penultimate time: *You Are Nowhere*, by Andrew Schneider, Kentridge's *Refuse the Hour*, and finally two plays that include children, *Before Your Very Eyes*, by Gob Squad, and *On the Concept of the Face: Regarding the Son of God*, by Romeo Castellucci. These performances with children are placed at the end of the chapter because although they also have to do with a repudiation of death, the use of the children places the future, futurity, into the center of our consideration. The chapter to follow this, on kairos, thinks against the future, and so the children in these performances form a bridge to thinking against the future as much as to thinking against death. Part of the work of this chapter will be to see how children in these performances distress and disorder the temporal determinations formed by both death and the future.

There are threads throughout the chapter on physics, and Einstein's thought experiments on time, including Bruno Latour's fascinating re-do of the traveling twins, a thought experiment that figures largely for Kentridge, and threads about railroads, in the thought experiment on the passing train that figures largely in Schneider, and in Kentridge's colonial railways in which Latour's figure of the mediator comes back into play. There are threads about falls, falling, about the self or the subject as relied upon to keep time unified; there are threads about theatre's control of linear time, processional time, and its manufacture of a time inside that time, an operational time out of sync with death and the procession to the black hole. It is a contention of this chapter, and of course this book, that the initiation of history in the ways that I mean it here depends upon this repudiation of death as the prevailing logic of theatre, of life, and of history. The interior and destination of chronological time and its manifestation as processional history is death. Always, processional history leaves the dead in its wake and offers only death, tick tock tick tock, in the futures it passes by and leaves behind, in the form of more dead bodies, the vanquished. But these theatre pieces show us how to think against that finitude, to think against death in order to found a politics of existence. Agamben's operational time, at work in the penultimate time before death, is a politics of time which becomes a politics of existence against finitude.

If my reader will bear with me, before we begin with the performances, I want to detail those concepts of Agamben's that will be central to this chapter. It is important as we begin to mark the difference between the following.

1. "The time *in which* we are."[5] That is, we are situated within and as part of passing time. This produces a feeling of helplessness with regard to the passing of time that is carrying us along with it; we detach and see ourselves being carried along with it like spectators. This can be understood in colloquial terms. We see ourselves age, caught up in our work lives, or our out-of-work lives, or our precarious lives, as they march forward. We feel ourselves as lacking a certain agency in the seemingly inevitable flow of it all. We talk about ourselves as surrendered to this flow of time, as time accrues in the form of age and exhaustion until the endpoint of our death. (Another day, another dollar.) This time is experienced as flowing along so that there is never any time left; there is no additional time. We look back and see that somehow, as our time flowed forward, we used up all our time. We feel slightly guilty, as if we might have done something differently and gotten hold of that time and used it better. But we feel that we did not and really could not have had any hand in it. This is our representation of chronological time. This is the representation of time with which our lives mostly concur.

and

2. "The time *that* we are."[6] This time is what Agamben calls operational. Operational time holds or refers to the construction of time, the time in which we make time. The "time-image"[7] that represents chronology is a spatial image of unidirectional linearity from past into the future. Operational time, by contrast, directs

> our attention to the way in which we can and do make time in a way that is out of sync with this representation of time. It refers to the way in which we can feel, think, and experience a time that is "additional" to chronological time, that prevents us from coinciding with the time images of chronology. This out-of-sync time, this additional time, is not something that comes from the outside of chronological time but is actually inside it. Agamben calls this time "messianic time"[8] as a way to think about how this time is created and how it functions.

The messianic should not be confused, as it usually is, with eschatological time or end of time. The arrival of the messiah does not announce the end point of time. Instead, it is a contracted time (in that it is limited), which is, in Paul's language, "ho nyn kairos"[9] or "the time of the now,"[10] which lasts until "parousia" or the end of time.[11] That is, the end is not a simple point, a last day. There is a time that time takes to end, a kind of elongation of time between the beginning of its end and the last day, the final point. Agamben calls it penultimate time. We can also translate this into colloquial terms. Between the day of the announcement of a terminal cancer and the actual death, there is a time remaining before that point, "the time that remains between time and its end."[12] And while the death will occur and the time before it is in one sense bound to a chronological representation of time, this time of the end does not need to be subordinate to chronological time. It can instead be a "transformative contraction."[13]

Agamben asks how we are to understand the "experience of the time that remains,"[14] the time of this transformative contraction, penultimate time. We can think of that time, he says, drawing from Gustave Guillaume, as operational time, a time that simply cannot be understood by the unidirectional line from the past to the future. Operational time is the time it takes to make a time image. The concept is a tool that allows us to see that time images do not exist independently, outside, already formed, as the unidirectional straight line implies. It takes time to make them, and the image in some way includes this time of the making. The time image thus becomes three-dimensional rather than only linear because it has to include an extra time, a creative time, a kind of balloon of time existing in chronological time that changes the linear pathway of chronological time. We will see shortly how the theatre makes time images and is itself the time of their making. Guillaume calls this three-dimensional time, which is a new representation of time, "chronogenetic time."[15] I find this an especially useful word for what theatre can do with time. A time image which includes the extra time for the making of time, the taking hold of time, is chronogenetic.

Agamben concludes that every time we represent time we have also installed in it "another time . . . that is not entirely consumed by representation [of chronological time]." Messianic time, an operational time, is the time we take to bring our representation of time (chronology) to an end and to create a different time image, "working and transforming" inside chronological time. He refers to this as "taking hold"[16] of time, of making time our time. This is differentiated from, as I have said above, a sense of having no say in the time that carries us along with it. We can mess with that time, we can prepare revolutionary time, we can, above all, remove ourselves from a procession to death, from definition by finitude. We can initiate time out of sync from processional history. The messianic, the messiah as the beginning of the penultimate, the beginning of operational time (the cancer diagnosis or the onset of a revolution, everything that is an inception and will have an end), is a way to mark the idea of "the time that we have." We understand this colloquially when we say, "In the time that I have left I will . . ." Time will not go on and on eternally. The inception, the messianic moment, inherently announces an end, but also that the time before the end can be one of temporal creativity. It is an announcement of having time in our hands to fashion time, and therefore existence, differently. The measure of existence is not death, or finitude. It is the time we have to change time. To change time is to initiate history, to nominate, by means of a temporal disorder, a new possibility for what it means to live.

You Are Nowhere

Andrew Schneider's *You Are Nowhere* works back from an end, a death, or a near-death experience; or it takes advantage of the temporal disorientation of the fall, the experience of the fall, to open the time before the end. *You Are Nowhere* works in a penultimate time, initiated by the accident of the fall from the temporal authority and verifiability.

This is a performance that it is impossible to write about, although of course I will try to do so here. This is because it quite literally changes time. It uses the capacities of theatre to change time, at the same time as it illustrates a battle against theatre's conventional time regimes, especially as exemplified in the stage manager—whose job it is, no matter the type of performance, to keep things moving along. The linearity of writing cannot possibly approach it. It inhabits and creates a different time. It is, as I overheard one audience person say as I was leaving the theatre, "a *total* mind-fuck." I could literally feel my mind and body going through a tectonic shift as I tried to comprehend what had happened when, as I will describe, another present opened up, in addition to the present I had been watching. The piece is full of meditations on time problems in physics, on simultaneity, on non-synchronicity or synchronicity, on breaking free of or being broken from the pattern of believing that one thing happens after another. It is full of death and what happens before it, all embedded in a theatrical apparatus so innovative, so time changing, as to stun us, absolutely change us.

The text of the piece was published in *Theater* in 2016. The text, as much as it tries to give a feeling for the piece—with highly innovative markings, cross outs, symbols, smaller font/bigger font, text written backward, upside down, letters extended, speech backward, in slow motion, indications of the electrical frequencies of the sound—remains only a chronological skeleton of a piece that sensorially, intellectually, and emotionally confounded chronological sequencing. After all, one of the lines of the play is: "You're really going to have to stop thinking of time as one thing after another though. That's the first thing you're going to want to try to change."[17]

The first two times I saw the piece were in January 2015 at a tiny performance space in Brooklyn: The Invisible Dog Art Center. I saw it a third time at the 3LD Art and Technology Center in Manhattan in the spring of 2016. The performance in this space was almost identical to the one at Invisible Dog, but I will describe what I saw at the latter. The space was white, cramped, low-ceilinged, rough-edged, and low tech. It was clear that there was going to be technology in the performance, because there were machines in place, but these were mounted to plywood platforms, secured by woven straps, low budget. There was a head-sized empty metal frame stage center. It was part of the alternating and precise play of lights to come; it illuminated at times with a blinding fluorescent glow, sometimes framing Schneider's face. The tension between the homeliness of the space along with the vulnerability of Schneider's body and the technical feats Schneider deployed was, somehow, part of how he accomplished what he did. It was a show that relied upon technology, some of it invented and manufactured by Schneider himself, but in which technology somehow disappeared as a technique and became constituent of temporal malfunction and disorientation. That is, this new time could not have appeared without it, but the new time was not about it.

Schneider arrives on stage, bare-chested, two strange small packs taped to each bicep with wires extending from them to his head, possibly into his headphones. He also wears a small microphone of some kind, even though he uses a wireless hand-held mic in much of the piece. In an alternating sequence of buzzing, like the horrible buzzes of industrial timekeepers in factories, alternations of light go out and on, and his own breath is out of synch with a voiceover of his own breath. During this, according to my notes, he is continually jolted, like being electrocuted, jolted backward, barely finding his balance, his own voiceover disjunctive from his live voice. The lights are bright, fluorescent bright, no warmth. The scuzzy abrasive sounds are loud and disorienting, the word "salad inserts" tell us that meaning is jumbled, disassembled.

With each extremely fast lights out and lights on, Schneider with absolute exactitude has appeared in another place on stage. His timing is so exact it is almost unbelievable. One wonders how a stage manager can call the cues for the lights that well, how the stage manager can see where Schneider has ended up in order to call the cue—it is pitch black when it goes dark. One wonders if he is an auto actor. He keeps touching those boxes on his arm. Is he in effect calling his own cues? His physical ability is dizzying, exciting, and, again, disorienting. He begins to talk, and it begins to seem as if he is in conflict with the stage manager, that person in charge of enforcing the time and timing of the performance, for ensuring that no matter how aesthetically radical the show may be, it moves successfully forward from beginning to end. It seems that it is they who are zapping him as he begins to try to speak. He tries, "Hey so we should really talk some time soon about what's happening."[18] He tries again after breathing, out of sync with his own voiceover breathing. He lip-syncs a song, "Lonesome Town." It is not clear yet, but it is setting up love, or the absence of it, which will become one of his themes. In the word "salad"/"buzzing"/"lights on–off," Schneider trying to speak, there is the feeling of mad current exchanges and zaps, Schneider going off-balance. The lights go out, they come up, and he is in an entirely different place on stage again. How did he get there so fast, without us seeing anything, what is happening?

He falls. Or when the lights come up, he is lying on the floor. Diagonally across the stage from him, in an instant's glimpse, like an afterimage, there he is, in the flesh, standing there. What was that? How was that possible? As the performance continues, blinding, ear blasting, incomprehensible, sometimes slow motion, the feeling is not one of continuousness. It is breaking, it breaks, it shatters, it is what keeps something happening, it is a dynamo, a force. Schneider appears disoriented. He is back up on his feet. He seems to collect himself. He tries an opening salvo. He mixes things up, his speech jumbled. He seems to be making an account of the fall, from somewhere outside himself, a scene with leaves, an un-showered body, paramedics who "begin to pick the gravel out of your face." He keeps trying to center himself: "Ok good so now we know right where we are right in the center with the past behind us and the future in front." The statement is obscured in static. He appears to be listening to words in his headset. He seems in distress and increasingly angry. He keeps looking up, as if in the direction of the booth; "I'm being told we have a limited amount of time here." He says, "We're already at the last act—the climax—the uh let's fucking go we don't have a lot of time here! But with the time we do have—and why don't we just keep track of that."[19] The lights turn blue. He begins to talk about the twelve steps of Alcoholics Anonymous, which are written in the past tense "as if you've already done them."[20] He is angry. The dissonance and disturbance and buzzing keeps pulling at him, keeps jolting him, rubbing out words that he mistakes, distorting his words, as if to get him back on track. There is a sense of the theatre apparatus both insisting on its timely procedures and also going haywire, as if some force were pushing the wrong buttons, turning the wrong knobs. He keeps getting disoriented, jumbling; something has happened, is happening, or has already happened—he declares that this is just a rehearsal "as obviously this has been extracted from a much shorter piece."[21]

As such, from the very beginning (which is already almost at the end), all representations of time made from linear, chronological, before/after time are already in crisis. The stage manager/crew/Bobby is unable to keep the Schneider persona within what was supposed to be the right story, the right sequence, in the center with the past in the past and the future in the future. That is, the time out of which chronological representations of time are made is already decomposed, so that this time inside chronological time is open and operative. There is no orientation with regard to the representations of time to which we are accustomed. This may be one of the foundations of the shockingly effective remaking of time in Schneider's piece.

Schneider stages the fall, the same fall, his fall, four times in the course of the hour-long piece. The fall is an image, and an experience, that presages, predicts, stands for, the means of cracking time as the before/after. It is an inception that marks the beginning of operational time. In the fall he has died, or he has had a near-death experience. It is impossible to verify which of these it is. It is in this ambiguity that death is not that which is the end, but that which opens both backward and forward, a distension of time; it is, as Agamben says, "an *already* that is also a *not yet*, a delay that does not put off until later but, instead, [is] a disconnection within the present moment that allows us to grasp time."[22] The fall as inception is also a disorientation, an escape from the authority of the verifiable, which is therefore also an escape from temporal authority, that of the stage manager, that of chronology. Constitutive of the experience of the fall is an interior time that is not "exhausted" by customary representations of time. From a slightly different angle, Schneider is no longer what philosopher Simon During calls the I of historicism, the transcendental subject whose job, since Kant, is to "unify the fabric of experience by getting time right."[23] Instead, he is a disassembling being opened to all the time there is in addition to chronological time.

Here it is clear that the stretch between the fall and death, or from death back to the fall, that penultimate stretch, is productive of thought, rather than thought's end. Schneider (without mentioning the source) quotes David Foster Wallace at length on what we can call penultimate time, operational time. He describes, from Wallace, "the literally immeasurable instant between the impact and death, just as you start forward to meet the wheel at a rate no belt ever made could restrain." This is an inception of thought that, freed of predictable time, marvels in, is terrified by, welcomes in itself a new capacity. Schneider via Wallace says, "If time is really passing, how fast does it go? At what rate does the present change? . . . What if there's really no movement at all? What if this is all unfolding in the one flash you call the present?"[24]

In *You Are Nowhere*, death gets reset. At the second staging of the fall, Schneider flies through the air and ends up in what the stage directions now specify as the hit-and-run position. We think of the story of the paramedics working on his face. We think of the story he just told of not being able to forget the sound of a man getting hit by a car, a man who may have lived or died, he does not know. He gets up and asks for a reset. But this happens after the coup d'état of the piece, which I will now try to describe.

At the brink of this stage miracle, in the midst of a scene furiously active, a mix of fragments and frames, he tells a story from childhood when he was with some boys in one of their houses. They heard a crash upstairs, a crash to the floor. They knew, just knew, that it was the boy's father who had just dropped to the floor, dead. He keeps going, through continuing glitches. He apologizes; "all this is obviously just sort of warming up—the whole thing won't be like this,"[25] and follows with a long speech from which I excerpt at some length:

> Oh . . . uh sorry got distracted I was supposed to be telling you a different story . . . but I want to start with a question. Uh . . . do you guys ever think that anytime you happen to think about or get scared of or have a really really close call with dying, that just right there in that moment, you actually already have died? And now are already dead. And that time just—there—just now—split off into a new universe . . . . And that there's all these alternate yous and alternate times spinning off and splintering and moving forward concurrently . . . and eventually there will be enough splinters and branches and spin-off, that literally every moment of your life, and now death, can be represented in the same "frame" the same "slide" . . . [*glitch/a collapse*][26]

This last stage direction indicates the abrupt and completely unexpected collapse of what we thought was the back wall of the performance space, used up until now for projections. Spectacularly, this curtain, as we now know it to be, crashes to the floor. It is a spectacle that reminds me of when Baz Kershaw says spectacle is "a sudden gap opening up between different ontologies or versions of the real, a kind of fissure in the way that knowledge of the world is usually assembled."[27] But what constituted the fullness of this ontology-changing spectacle was what was on the other side: an audience identical to us, sitting in front of an identical booth, with an identical frame, watching an identical performer performing an identical show. The image was so precise and perfect that almost across the boards most audience people thought that the other audience was a mirror of themselves, and spent some seconds attempting to locate themselves in it.

Gradually, very gradually, like our own coming to consciousness after a fall, it becomes clear that this is a simultaneous reality, that on the other side there had (perhaps, although we did not hear it) been a simultaneous event and that, as Schneider says early in the piece, concluding a short video presentation about the simultaneous lightning bolt and the relativity of perception with respect to motion, "we exist in each other's realities, but maybe not in the way we think we do."[28] There has been no indication of a second audience going into a second space in the theatre prior to the show, not a sound from the other side of the wall/curtain, not a single rustle or cough. And yet, there they were, as if in a time split, a time shift, an ultimate dis-coordination of time as we normally experience it. I saw now that the identical performer, Schneider's doppelgänger (double), split self, other self, alternate self, was the Schneider illuminated so briefly at the onset of the piece while standing diagonally across the stage from Schneider's prone "hit-and-run" body. We are suddenly residents of a void that has appeared in the present. Not only Schneider but we, the audience, also have these alternate selves and alternate times. We are encountering a spectacular discontinuity of time within the present, that we may or may not have been sharing with the other audience. It is really very hard, in this spectacular moment, to understand the nature of the time, the present, that we are in at all. It is as if, as Schneider says, our brains "can't really work like that."[29]

Schneider appears now to have two selves. One is A and the other is B, in the script. The exactitude and precision with which he and his physically not identical but very closely alike doppelgänger begin to speak and move is breathtaking. It is like an exquisitely practised mirror exercise, except that they frequently execute prolonged movements in which they are not looking at each other and, far more than any chorus dancer, it is like they are the same person. They list, identically, in great detail, random facts about themselves and their lives, a range and depth of things no one else could know. And yet, it is as if they wish to be free of each other, they wish they could get out of sync, sometimes it seems that they are trying to trick one another up. They say,

> **A/B** Whoa. Just. Fuck whoa. What the fuck man? ok. ok. O. Kayyy—how do we get out of this. We can work together—there has to be a way. It's obvious that no one has any idea what the fuck is going on . . . we must have been offset or out of phase or not always and eternally doing the same fucking thing and saying the same fucking thing at the same fucking time what is this?[30]

After a long exchange of life details, spoken in sync, each appearing to have the same history, there is finally a split. A speaks, and he speaks alone. Before long B joins back in. They appear to be getting more and more furious with each other. There is a loud explosive sound. This is the second fall, the same hit-and-run position as at the beginning. B watches A without emotion. It is the same fall as when we saw B as Schneider's other self for the first time, ever so briefly, as Schneider was "beginning" the show, watching Schneider on the ground. We are back there again. This is the moment of the reset.

Schneider gets up and, pulling the rigging from which the curtain dropped down to stage level, begins to reattach hooks to the edge of the curtain, preparing to restore it to what was at the "beginning" of the show, as if we are going to do the death again and again. But just here something different happens. House lights come on, and the two talk to each other as if they were simply actors, asking each other where they are from. But, curiously, B begins to act as if he were a mentor, as if he has been around the block more than A. It begins to seem, only intuitively, only ambiguously, that B is the other self of the fallen Schneider, the part, perhaps, on the other side of death. It begins to seem as if B can see, from the point of view of Schneider's death already having happened, A's penultimate time.

B asks A, "How many times in your life do you think you've almost died." Schneider tries to respond: "Uh . . . I don't know a lot. But everyone almost dies a lot. I mean that's what life is, right—like a big long series of almost dying, until/." B interrupts: "Physics works just as well backwards as it does forwards. The trouble is no one has been able to figure out how to make it do that."[31] It is as if we can move backward from death. We are, repeatedly, in this piece, moving backward from death. And here there is backward speech, something being said backward over and over. A gets a bloody nose. More backward speech. A is again fallen, on the floor in the same position. The sound, used throughout the show and now here again, is a "10K tinnitus tone"; tinnitus is a chronic ringing in the ears. It is maybe, here, the sound of death.

The house lights come up. A and B are now facing each other's audience. B asks the audiences to move to the other side. Puzzled, we begin to pick up our things and do as we are told. We step around or over the curtain that is still in a long heap on the stage floor. When the switch has been accomplished, B begins talking to A in an aside in what sounds like a description of what it will feel like when A dies. He follows this with the line I have already quoted: "You're really going to have to stop thinking of time as one thing after another though. That's the first thing you're going to want to try to change . . . it's not that easy to understand. At least in the way that alive people understand things."[32] B can understand what "alive people" cannot. A self that splintered off at the point of the fall, a self that was, what, always there, a double in a parallel universe, who splits off at the point of the fall, the voice of the one who sees the other in the fall, in the penultimate time, both inhabiting this creative delay, the already but not yet. The curtain goes back up, slowly. B and his audience slowly disappear behind it.

There is machine gun fire, a short segment with glitch after glitch interspersed with A gradually eliminating himself from language, or going backward from language. The theatre is pulling back on itself. Back from the end. The lights go to blue and the curtain collapses again. Another mind-blowing shock. B, the other audience, their seats, the set are gone, as if up in smoke. Just vanished as if they had never been there. It is very dim. It seems smoky. A beautiful "last" scene begins in which A is speaking backward. A match is struck, an incandescent bulb on a lamppost lights up, the first warm light, perhaps the lamppost outside the deli where Schneider fell. He keeps talking backward, the lullaby "Golden Slumbers" plays, the bulb smashes to the floor, lights go out and point by point. In the stage directions Schneider indicates that the many tiny LEDs that begin to dot the ceiling are stars. Mysterious. Unknowable. There we sit, trying to come back to our bodies, and into the time that we must now go back to inhabiting. It is hard. People linger for a very long time.

In *Infancy and History: On the Destruction of Experience*, Agamben writes about two accounts of falls. In these stories, as well as in Schneider's fall, there is a radical displacement of self from self at the limit point of death, of near-death experiences, with selves appearing as differing temporal experiences. There is an elimination of linear temporal orientation. These falls, as well as Schneider's, could be described in During's terms as "the disfiguration or destitution of the subject as a general instance in charge of coordinating the dimensions of experience starting with the synchronization of a diversity of heterogeneous durations within a unified now."[33] Schneider's image of this dissolution is initially summoned as an image of simultaneity with the other self. At the point of the fall, the second self appears as a nightmare of unity of time. But it is important to differentiate between simultaneity and synchronization. They are not simultaneous, because we know that simultaneity is relative; it is elastic, its perception relative to the observer. Instead, the two selves are synchronized. Synchronization, unlike simultaneity, depends on being in time, adjusted to a clock, obedient to time, just as Schneider has been forced to try to do throughout the piece, especially with his lip-syncing voiceovers. At his fall, his near death, he is in a time in which he can engage in a heated struggle to achieve separation, disunity, to go out of sync. The struggle for disobedience from unifying time is staged. We can think here of Groucho Marx, in the mirror scene in *Duck Soup*. In a more mischievous way than Schneider, he is trying to outwit his other self, as they both traverse the space of the open door between them with identical gestures, miraculously appearing with the same hat even though they could not have known the other was putting on a hat. Gradually Groucho "wins" his independence from this temporal unity, this "in-sync." Groucho, that perennially uncatchable figure of un-doing, has to be a precedent for Schneider here. Schneider and the other self, the doppelgänger, also, finally, go out of sync, and it is here that Schneider's fall echoes the falls that Agamben cites.

The first is one that the Renaissance essayist Montaigne describes. An over-zealous young horseman rode straight into the path where Montaigne sat on his horse, knocking both he and his horse over backward, "so that the nag lay along astonied in one place, and I in a trance groveling on the ground ten or twelve paces wide of him; my face all torn and bruised . . . with no more motion or sense in me than a stocke." He goes on to describe the strange sensation of knowing himself almost already dead and feeling a separation of "my selfe go from my selfe." He speaks of being barely able to see, his vision perceiving only a surrounding dimness, and

> I perceived myself all bloudy; for my doublet was all sullied with the bloud I had cast . . . . Me thought my selfe had no other hold of me but of my lips ends. I closed mine eyes to help . . . to send it forth, and tooke a kinde of pleasure to linger and languishingly to let my selfe go from my selfe.[34]

He is at death, split by death, aware only that he closes his eyes and lets his breath go from his opened lips to be dead. But then he goes backward, as he was not yet quite dead, and finds in the penultimate that he is going through the pleasurable sensation of falling asleep. It is not hard to see here a parallel with the experience Schneider is staging, the one self watching the bloodied prone self, or the sensation of being in the body, an opened lip away from death, the sensation of lingering there, asleep, or as if asleep. And then, in Montaigne's account, he comes back into a single self in an almost unanticipated return, time reversing itself but doing so through an opened dwelling of and in pleasure not constituted by linear time.

The second fall is narrated by Rousseau. His was caused by a charging Great Dane who was running at top speed straight at him. Rousseau describes his plan to leap over the dog just as it arrived in order to avoid the fall. But this was

> my last thought before I went down. I felt neither the impact nor my fall, nor indeed anything else until I eventually came to. The first sensation was of delight. I was conscious of nothing else. In this instant I was born again . . . . Entirely taken up by the present, I could remember nothing; I had no distinct notion of myself as a person, nor had I the least idea of what had just happened to me. I did not know who I was, nor where I was; I felt neither pain, fear, nor anxiety. I watched my blood flowing as I might have watched a stream, without even thinking that the blood had anything to do with me.[35]

Rousseau experiences the fall initially as a death, a nothingness. He returns to a form of consciousness in which the self, the I as a unifier of time, does not seem to exist. The temporal locatedness of such a subject in the before and after of chronology has gone absolutely missing. For Rousseau this feels like a birth, which he specifically describes as a curiously non-temporal and therefore blissful state of being a watcher of his own body, but without knowing himself as the watcher and without recognizing himself in the blood. In Kantian terms, both the transcendent and sensible disappeared as an apparatus of knowledge production and temporal coordination. It is an unverifiable anomaly, a no knowledge, an inception.

Another story of a fall that is directly related to temporal inceptions is from the brink of the twentieth century. It is an account by historian and Harvard professor Henry Adams of a fall he took (perhaps metaphorical) at the Great Exposition of 1900. It is included in the essay "The Dynamo and the Virgin," in his book *The Education of Henry Adams*. During his visit to the Exposition, he is guided by the inventor Langley, who is interested only in showing Adams the newest scientific work, what Adams calls "force." He shows Adams "the great hall of dynamos," with the giant machines incomprehensibly producing electricity out of the heat from the coal hidden off-site. Adams is captivated by the relation between the dynamo inside and the coal-run engine house outside. He cannot grasp electricity as fact; he cannot connect the steam from the engine house and the electric current. For him, this "break in continuity amounted to an abysmal fracture in the historian's objects" so that he could only see "an absolute *fiat* in electricity as faith." He describes the work to which he has devoted himself: "Historians undertake to arrange sequences—called stories, or histories,—assuming in silence a relation of cause and effect." He says that he "had even published a dozen volumes of American history for no other purpose than to satisfy himself whether, by severest process of stating, with the least possible comment, such facts as seemed sure, in such order as seemed rigorously consequent, he could fix for a familiar moment a necessary sequence of human movement." But he found himself believing that "the sequence of men led to nothing" and that "the mere sequence of time was artificial." He turned instead to "the sequence of force," and it "happened" then that

> he found himself lying in the Gallery of Machines at the Great Exposition of 1900, his "historical neck broken by the sudden irruption of forces totally new." He concludes, after this fall, that he "must treat [the forces] as they had been felt; as convertible, reversible, interchangeable attractions on thought" and that "here opened another totally new education, which promised to be by far the most hazardous of all."[36]

Adams the professor has for years been obsessed by knowledge and verifiability, hoping that by sticking to the facts laid out in sequential order history will be legible, appraisable, with cause and effect measurable. But he is overtaken by force, falls hard, tripped by the "abysmal fracture," a void in the heretofore apparent unity of time carefully constructed by craftsmen such as he ("the mere sequence of time was artificial"), a representation of time created from the time of chronology, which does not allow for "the forces," or what we might call the additional time within it. With his neck broken by a new historical force, lying on the floor, he experiences his renewal as the charge of forces that are "convertible, reversible, interchangeable attractions on thought," a situation of jumbling, forward/backward, converting in thought, as opposed to thought that plods obediently, sequentially along.

All four falls liquidate that subject expected to "unify time" by "getting it right" and, by doing so, to remain subjected to the politics of time that attempts to convince us that we are capable only of this operation, and not of operational time. The fall—the fracture that seems to initiate operational, penultimate time—inaugurates a "hazardous" education, as if there is nothing safe in opening new temporalities, in looking for them, to the extent of breaking our necks in the effort.

It seems that, in fact, Schneider sets out in search of this "hazardous" education in a deliberate act of friction and defiance against the sequencing of dominant time. He throws himself into the midst of the multiple entities created by physics, none of which cohere either. He has no choice but to comply with theatre's "timing," the compulsion to get the cues right, the exactitude of the synchronized voiceover, the starting over if he misses a cue. But he struggles to escape it, a painful, frustrating, infuriating process, with the glitches and the static, and the rubbing out of his face to black, and the stage manager's hurling CDs at him from offstage. He has to go really fast. He only has an hour. There is a digital time signature moving furiously at the bottom stage left edge of the curtain/wall. He is on a clock.

In addition to theatre's sequential time, there is another imposition of sequential time. This is the twelve steps of Alcoholics Anonymous. The near-death/death, all those times he had a near-death experience, seem to be associated with drinking. There are allusions to his drinking throughout the show. Waking up in the hospital covered in his own vomit seems like an alcoholic scene. His preoccupation throughout the piece with love, its failures—particularly with the girl who, he tells B, did not come to the show tonight—seem linked. His explorations of the dilemmas of time are mixed, always, with the problem of not ever being able to know what another is feeling exactly, not being able to know it as if you were them, are interwoven into physics. His voiceover of a YouTube video illustrating Einstein's relativity, via the example of a fast-moving train with its passenger and the stationary person on the platform, deviates into his own meditation on whether the two might have gotten together, if after seeing her on the train he tried to reach her by throwing a message into the void of Craigslist. So, perhaps, he has come to AA. His life is a mess. He keeps having near-death experiences. He thinks about death all the time.

Almost immediately, after the first frenzied exchanges with the theatre apparatus that is trying to get him in line, it seems that he may be addressing an AA meeting. He comes to the meeting with notes; he thinks he remembers his notes. (People embarking on Step Four are encouraged to carry a notebook around with them to record their thoughts.) But it immediately devolves into a mockery of Step Four: "Make a fearless, searching moral inventory of yourself."[37] He mocks this idea and then there is a gunshot.

He breaks off into a nearly inaudible pleading, as if trying to communicate quickly and in a panic what he really wants to say, perhaps to the person, perhaps to the girl, for whom he has agreed to attend AA and try to obey the awful map. He tells her that he is trying, really trying and that he is doing it for her, but somehow nothing is as you thought you knew it.

There is then a furious spat with his own recorded voice. All this is going very fast, the frame lighting up white then going to dark over and over, every one, two, three seconds, entirely discontinuous. Then, as front LED lights are activated, Schneider seems reconnected to the message he is supposed to say, the map, the appreciation for the moral inventory map where all the lines are straight.

In *The Alcoholism and Drug Abuse Client Workbook*, Robert R. Perkinson says:

> Going through life is like going on a long journey. You have a map given to you by your parents . . . . The map shows the way to be happy . . . if you make wrong turns you end up unhappy and addicted . . . . What you need is a new map . . . . Twelve-step programs give you a new map. It puts up 12 signposts to show the way.[38]

This is perhaps the core from the real world in *You Are Nowhere* that stands for a very real kind of domination by linear time: those people standing up in AA meetings, with their notes, trying hard to follow the signposts on the map on their journey. It is a politics of time, a morality of time, that is unanswerable to the explosive unverifiability, unmappability of experience, of what happens in the inception of operational time, the hazardous journey, even as an alcoholic suffering near-death experiences repeatedly, to open time, to initiate time. I have no idea if Andrew Schneider is writing from his own experience or not. It does not matter. The desire to escape the disciplines of dominant representations and practices of time, and to reject these in the theatre and turn to theatre's other temporal potential, finds an illustration in this situation in which death is always close, but is not, it turns out, quite the end we thought it was.

Refuse the Hour

> O Death, where is thou Sting?
> O Grave, thou Victory?
> O Death, where is thou Sting?
> O Grave, thou Victory?[39]

This version of the famous passage from Corinthians is at the heart of the chamber opera *Refuse the Hour* by the South African artist William Kentridge, which I saw at the Brooklyn Academy of Music in 2015. The opera is another staging of the penultimate: an affirmative, exuberant performance. It ends simultaneously with one of Kentridge's signature shadow processions on an upstage screen moving stage right to left, approaching the black hole, the grave, death, *and* a procession of the live performers and musicians processing downstage, toward us, singing, a cacophony of voices and instruments, a jubilant affirmation and refusal of the end of this night in the theatre. In a talk given during the MET exhibition, he says, in the full version of the quote with which I began the chapter, that the piece is "a celebration of making against the effects of our eventual disappearance. We're not going to escape our journey to the black hole at the end, however fast we dance and run away, but that dance and that run is still what it's all about."[40] In the extraordinarily beautiful book assembled from the installation, he writes, "Everybody knows that we are going to die; but the resistance to that pressure coming towards us is at the heart of the project. At the individual level, it was about resisting; not resisting mortality in the hope of trying to escape it, but trying to escape the pressure that it puts on us." The refusal of the end is linked to the colonial refusal "of the European sense of order imposed by time zones."[41] The refusal of time creates the condition of a penultimate time that is characterized, as Bruno Latour might say, by a profusion of entities and times, a place where, so to speak, the smooth path of the train (the monolithic colonial project dependent on setting times) is interrupted when the tracks are set on fire.

Kentridge also did a video installation created for dOCUMENTA 2012 called *The Refusal of Time* at the Metropolitan Museum of Art. In April 2012, working on the same material, he delivered *Six Drawing Lessons*, a series of one-hour lectures, as a Norton Lecture at Harvard University. These are online, accompanied by Kentridge's films and films of his work in his studio, and have also been printed in another beautiful book from Harvard University Press with the same title. My thinking on *Refuse the Hour* here is enriched by these video lectures, the printed text of them, the book from *The Refusal of Time*, and a pre-performance interview at BAM with Kentridge and his dramaturg, the Harvard historian of science and physics, Peter Gallison. The principal collaborators for both the installation and performance (among a large group of them) are, in addition to Gallison and Kentridge, South Africans Philip Miller as composer (who has released an album/soundtrack, *CountingTime*) and Dada Masilo, the extraordinary black South African choreographer and dancer.

The piece is built around seven short segments of text by Kentridge, performing in his invariable white shirt and black trousers. I will call it a monologue, even though it does not feel like one because the bits of text occur associatively and relationally in the fabric of the piece as a whole, the opera, the singing, the dancing, the musicians. In a sense, he positions himself as one entity among many. He speaks of the paradox of the twins, of Perseus and fate, space as universal archive of images, the studio, resistance to entropy, stopping time, reassembling the fragments, photography and turning time into stone, the European synchronization of clocks, the spread of the meridian through the colonies of Africa, the returned twin, and what happens on the edge of the black hole. Around him on the stage, many object entities, many people, musicians, singers; Dada, tiny, head shaven, taut and iron-muscled, Kentridge's partner in thought and movement; Ann Masina, black South African gospel, choir, and opera singer, magnificent, a maybe 400-pound force, who sings the song of colonial revolt; white Australian genre-breaking experimental vocalist and opera singer Joanna Dudley, white-skinned, helmeted black hair, red lips, expressionist vamp (also Lulu in Kentridge's production of the same), who sings backward, pulling back the words, swallowing them; Thato Motihaolwa, a black South African actor, short in a yellow shirt, who gathers, punctuates, arranges time, reminds us of the time, calling out, gathering. Large silver cones, reminiscent of megaphones, are a central visual component of the piece. They are used to send out and amplify the voice. They are the breath, the voice, as signal, sent forth.

The musicians play an assortment of instruments, many of them dependent on breath and air, brass instruments, tuba, trombone, bellows with horns, accordions, breathing tubes. The artists worked with the human heart and breath as clock, beating, pumping air. They say, "with each breath, we pump out images and transmit ourselves and traces of ourselves."[42] They were inspired by a giant system of "pipes and tubes . . . constructed with the idea of bringing hygienic time to institutions and citizens of Paris"[43] laid out under Paris in the mid-nineteenth century. They describe it as a "mother clock . . . with powerful bellows, and every minute the mother clock would give a pant of air, a breath, which would travel down the thick pipes in the grand boulevards," through them to all the smaller clocks so that "every minute the clocks would all shift one minute. Breath, wait a minute, breath . . . . A whole city breathing in unison, regulating themselves to the mistress clock, who stood in for the idea of perfect time, perfect order."[44] The breath and air-driven instruments, the breaths of the singers, arise in part from Kentridge and Gallison's fascination with this system for ordering time. They also worked from the late nineteenth century's telegraphs and signals and measuring and the development of synchronization of times across the globe, and then, in 1905, Einstein's theory of relativity; all of this influence other parts of the soundscape. They worked from old phonographs and the Edwardian hurdy-gurdy. In one gorgeous segment Dada, with large silver cones covering each arm and one leg, stands on a small circular platform that Kentridge, sitting on the floor, slowly turns with a lever. She dances the movement of a clock's hands. Time, as in Guillaume and Agamben, is turned to three dimensions. One coned arm crosses another, the cone on the foot intercepts; she is turning her own body too; it is all very slow, very precise, not a movement out of place. She is accompanied by the hurdy-gurdy. There are also the colonial dance bands of Kentridge's youth. There are automatically playing drums as part of an assemblage hung over the stage that plays the overture, taken from Berlioz's idea of a composer who could conduct across distances to far away musicians via "a telegraphically linked baton . . . a musical science fiction of time control."[45]

The musicians are at stage right, with an assemblage of instruments. Around the stage, some on moveable small platforms, there are assemblages made with metal rods and more cones, bicycle tire frames, like clocks moving, mounted on long tripods on wheels, bicycle chains working like pulleys, spinning, spinning, a large object made of wooden arms moving mechanically, an exaggerated and multi-directionally moving ruler with hinged arms, advancing. There are chairs. The silver cones rest on the floor when not in use. Dada, Massina, and Dudley are in Kentridge's signature costumes, part paper, imprinted with letters, words, sentences—paper as Kentridge's central medium, charcoal on white paper.

Most significantly, perhaps, there is a screen stretching across the entire back of the stage, onto which are projected films of Kentridge's studio: a place of transformation, performance, labor, experiment, over and over. There are words, grids, measurements, drawings, the images multi-layered, often split-framed, full of paper and torn bits of paper reassembling. Sometimes we see Kentridge working, the artist in his white studio, but deploying film so that his labor in the studio, drawing, moving, can be run backward. Sometimes the studio space becomes entirely a film. There are five films played, which I will describe in detail a little later. Each is a historical, actual scene of time making that is re-made, revolted against. Kentridge works in part from the aesthetics of the magician and early filmmaker George Méliès. He works in film because film has been the medium of time montage, time reversal, time succession, frame by frame. If in the nineteenth century, he says, "Photography turned time into stone," the moving image "made an examination of this congealed time possible."[46] Dada is turned to a dancing paper silhouette of herself dancing live onstage. Sometimes every entity in the films is made of paper, sometimes a live body. The studio, always in sight as we watch the live performance, seems like the continually generative heart of the stage and what happens on it. It is as if the nothing on stage has been prepared in advance to be laid out, performed, in its sturdy chronological order. Instead, the stage is always in interaction with the creativity of the studio, the work on changing time in the studio. It is the depth, the enlargement, the deformation in chronological time. And it is all of it, as Kentridge says, about the creativity that happens before death, the swerve away from the fated end, penultimate time. It is Kentridge's chronogenesis.

It is complicated to write about this piece. Its thought is associational, not linear. My attempt to write about it becomes associational also because there simply is no linear way to describe it. And so, because of the importance of the studio, its brilliant relationship to what is on stage, its profound thinking about time, I want to detail features of the studio from Kentridge's lectures on it before returning to the other complexities of *Refuse the Hour*'s time making. It is in the studio, especially, that *Refuse the Hour* initiates history, in the sense that I mean it throughout this book—which, to refresh my reader on the point, bears repeating. To initiate is to create a disorder by means of temporal innovation within what we commonly understand as history, and, by means of that disorder, to nominate a new possibility for what it means to live. And history is the manifestation of a radical human capacity, deployed along with or in the interest of other entities, and found in temporal interstices and innovations, in order to inaugurate universal (belonging to everyone and excluding no one) conditions of justice, recognition, and equal distribution in particular local and/or globally related situations. So, the studio.

In the studio, Kentridge says, he has the experience of stretching and expanding time (talk).

In the studio, "A rhythm of walking that stops time. Endlessly walking but staying deep in the studio."[47]

In the studio, the walking, two of him walking in opposite directions. "Trying to make sense of the idea of time and trying to find the material in the studio. Trying to find the transformation of different degrees of tension to squeeze an insight from the rock of stuck thought."[48]

In the studio, always in black and white as he works on *The Magic Flute*. The story of the enlightenment forced on Africa. The story of the Herero revolt, suppressed by the German colonizers in an act of genocide, skulls shipped back to Germany for measurement and display. Pinned to the walls: "The libretto of the opera / The music of Sarastro. / The Typewritten *Vernichtungsbefehl*. / The foot lashing of Monostatos. / The whipping of laborers in German South West Africa turned into postcards with Christmas greetings—an archive of colonial images"[49] He makes a miniature theatre in the studio. "The studio becomes a compression chamber for the images, ideas, historical links. [we see the multiplication of beings, entities and events] The miniature theater, a studio reduced further, a space for the elements to bounce against each other—the measuring of the skulls, a pair of dividers, a French nineteenth-century egg whisk to stand in for a Herero woman."[50] In his studio, he paces, circles, counts. He joins what Bruno Latour calls "time-producing collectivities."[51]

In the studio, crowded with entities, the linearity of language (the train, if you will) breaks down. It is not stream of consciousness, Kentridge says, because that too implies linear progression. Instead,

> there should be a way of registering the highway of consciousness. Many thoughts and different lanes, overtaking, pushing one thought onto the verge, becoming stuck at roadworks. Until one thought emerges ahead of the others, takes the off-ramp onto the page. This is the problem of putting elements of the lecture into a linear progression, a lateral procession, when what we need is a full frontal assault of all the images together.[52]

As at the end of *Refuse the Hour*: that joyous frontal assault, detoured from the lateral procession to the black hole. In the lecture in which he describes this, there is an image from his studio on the screen behind him. There, a notebook is open to two blank pages and there is Kentridge's hand, drawing with a single line a tangled, multi-directional, no-directional mess.

In the studio, in the miniature theatre he makes for *The Magic Flute*, he makes a film for representing the Herero revolt. The entities—a divider, an old French beater, charcoal drawings, shadow puppetry, of torn, reassembled paper figures in silhouette. In the film, there is a traveler, a configuration of paper bits, moving seamlessly along a little miniature rail line. There is no disruption. His body moves along at the same rate as the train. The traveler is oblivious to the scenes he is passing; vintage footage of chained Herero, and of trains, the colonial trains, oncoming, looming trains, and a blown-up train. In *Refuse the Hour* the dates of colonial revolts are projected in the studio and Kentridge names them, the demand to "resist the weight and control of Europe . . . . As if blowing up a train line could blow up the pendulum of the European clock, which swung over every head."[53]

In the studio, "Shaking the different elements as if the jolt of each stop is enough to shake them into a new configuration . . . . To try to undo entropy, to shake a box of letters and find a new sentence . . . walking around trying to resist the arrow of entropy, of disintegration."[54] The labor of the studio, in the mediation, the aging, the wounds, the multiple entities, the charcoal drawings that magically undo themselves and come back together again, shaking a box of letters, trying to undo entropy, comes on stage as "to unsay. To unremember. To unhappen"[55] and "a tear forward becomes a repair backward."[56] On stage, Kentridge meets Masina as fate by the side of the road. She sings "your time has come." Our time has come. But the fatality of the message is changing, because we are seeing unhappening, because our time no longer means death, necessarily.

In the studio Kentridge films himself pacing, frame by frame, stepping over chairs. Dada joins him in the studio. She steps over chairs, she carries him on her back, him draped over her back face up. He repeats, he keeps walking, he counts the time. This kind of hampered, obstructed, difficult, repetitive labor is necessary. He films himself going backward. He films himself sending out and receiving back—pitching a notebook forward, working and working to see exactly the position of the body so that when he reverses the film and it comes back to him it will be exactly right. He kicks his hat and it comes back.

> We look at something usually taken for granted, the passage of time, the movement of a person, and in the studio we demand its reconstruction, its shattering. Time changed into the marked graduations of animation: learning its grammar in the hope that in the end something different will emerge. The same happens in performance. The movement of the body, the backward and the forward walking, allowing it space, time to show us something unpredicted, unanticipated.[57]

In repetition, there will always be something new. In the repetition there will be a fracture, a difference, a crack—something is suddenly there that was not there before. There is always something new in the repetition. This is a temporal capacity of the theatre, this inner contradiction. It is an art form which depends upon repetition, but in this way it always shows an inception of some kind, however small, which changes the progression, the timing, of the repetition. Kentridge makes lists. He says:

> I find I repeat the list four or five times in different notebooks . . . each time I expect the list to be different; each time, to my surprise, it is the same, or almost the same. But in the re-ordering, the slight shift, the word that is illegible, we make some new crack, a new element enters the list, makes a space for itself—and this is the guest we have been waiting for.[58]

All this, then, is the rich material, the chronogenesis being labored at in the studio, which is sent out, returned, sent out again into the labor on the physical stage we are watching, where Kentridge, beginning the first section of his monologue, tells a story of being a boy on a train with his father. (Presumably, he and his father are on a train running on tracks built by a European power to implement the importation of settlers and the exportation of wealth from its colonies.) His father reads him the story of Perseus, with all its twists and turns leading to what was already fated, his throw of the discus that kills his grandfather. Kentridge literally jumps with agitation in a burst of energy crying out, "For me this was intolerable . . . . How could so many chance events . . . conspire to make the predicted inevitability?"[59] He cannot accept that every single decision Perseus (or therefore any of us) makes might lead only to the thing that was already there waiting for us.

It seems that this experience of the intolerable is a story about the inception of Kentridge's refusal of time and fate: that is, the time that leads inevitably to a fated end. In Drawing Lesson Six, "Anti-Entropy," Kentridge turns that discus throw into an image of return. The arc of the discus arches its course to a return to the sender, just as the story of Perseus, he says, is a "poem of return."[60]

The idea of return is threaded throughout the performance. He works constantly on the idea of sending signals out, through the universe, and how the signals return, how we send words out, and then they return, backward. He speaks of a film made in the studio in which his son throws a jar of paint all over the wall, tears up paper and throws the pieces everywhere. Then they run the film in reverse. Everything returns. His son is overjoyed that this can happen. Masina and Dudley pick up the cones. Masina, magnificent in yellow, is up in a box, Dudley is on stage, poised with exactitude on one stiletto heel. Each raises a cone to her mouth. Dudley sings out to Masina and then sings the words backward, pulling the words back, a curve of return that somehow moves past fate. It is an almost undoable feat, but she does it.

Because throughout *Refuse the Hour* Kentridge and Gallison are working with physics, with the history of the idea of time and its implementation, Einstein's travelers become for Kentridge another way to work with return, with changing time. Although his reference to the twin travelers comes near the end of the monologue, I want to address it at this point because, read with Andrew Schneider and Bruno Latour especially, it has everything to do with the process of chronogenesis and the initiation of history at the core of the piece. Let us begin by returning to Schneider. We'll travel for a while here before returning to Kentridge.

In an early section of *You Are Nowhere*, Schneider performs a voiceover to one of the most frequently viewed YouTube "explanations" of Einstein's theory of relativity, an example Einstein provided. We watch the graphic on the screen behind him as he does the voice. There is a moving train with a passenger in its center, and the person who is stationary on the platform watching the train go by. Two bolts of lightning hit, at the front and back of the train. The observer sees them as simultaneous, but the woman in the speeding train sees the bolt at the front of the train happening first. Both of them are correct, because "from different reference points there can never be agreement on the simultaneity of events."[61] The video is notable for its general incoherence as an explanatory device for physics novices, at which it is clearly aimed. It excises detailed parts of the problem in favor of a simplified portrait that cannot, thereby, account for the grist of the apparent conundrum or its solution. The animation is also notable for the absence of any detail in the world it pictures. The world is reduced to a flat green emptiness extending far into the distance, the palette for the simple gray path (no railroad ties, no bolts, no stones, no weeds, no trash) that lays down along its length. There is nothing else, except the edge of the platform, a simple square adjacent to the "track" that is similarly denuded of detail. There are no ticket machines, benches, newspaper stands, trash cans, schedules, no breeze. The man on the platform and the woman on the train have enough cartoon detail to conform to white heterosexual sexist norms but are otherwise stiff moving caricatures. How does one illustrate Einstein's train, or his rocket ship? What did Einstein see when he imagined the train in his example? Did he see a "pure" train as this YouTube example seems to try to achieve? Or did he see the trains of his time—perhaps the train station close to the patent office where he worked? Einstein's world was a world of trains, and not just for transportation. Did he see the trains of Kentridge and his father, trains for imperial expansion, for the colonial trade, bringing settlers in, goods out? Did he see the train's genocidal consequences, the implementation of manifest destiny, as in the United States? There the first train ran in 1829, and the railways, as in Africa, carried white settlers across the high plains where building the trains was decimating the buffalo and hence the plains Indians. The Indians attacked the lines, sometime by derailing them. Did he see the fugitive uses of the train, hobos "riding the rails," and the metaphoric underground railroad?[62]

In Schneider's piece, this "pure" graphic of Einstein's train in its smooth landscape along its ultra-smooth track is cut off by Schneider just prior to its conclusion. Schneider derails it, and derails its function. As I have already described briefly, he turns it into a love story, combines it with more recent physics into the possibility that the two people, who have such different perceptions and cannot experience the same thing simultaneously, can get together. The one on the train seems to be speeding forever across the bland and empty green, never to disembark. In a whoosh, she is gone. But in Schneider's story, the man on the platform has thrown off his function as featureless example because he has been smitten. He posts a missed connection to Craigslist. Now the woman too takes on features. She gets off the train and checks Craigslist and sees his post and her heart leaps. What, Schneider asks, were the chances of this, that someone "took the time to actually throw a post into the black hole of the internet never to be seen again"[63] and here she is? He has thickened the journey of the train and the world it is in.

In his essay "Trains of Thought: The Fifth Dimension of Time and its Fabrication," which I discussed briefly in Chapter 1, Bruno Latour stages the paradox of the twin travelers, Einstein's other most famous and notoriously difficult thought experiment, which Kentridge also uses. Instead of a rocket ship, as in Einstein, Latour situates one of the travelers on a journey by train, and renews, redoes, derails the reason why he, moving through the landscape so quickly, ages much more slowly than his twin, who is left on the ground. In this description, readers will understand my emphasis above on the smoothness and absence of detail in the YouTube demonstration of the train and its passenger. In Latour, the other twin, the one left behind, also sets off on a journey, but it is through a jungle, on foot. She has to cut her way through to make a trail. In the process, she is cut, wounded, as she herself wounds other entities. It is an "excruciating" trip that Latour proposes she will remember her whole life, especially because each tiny movement has been "a complicated 'negotiation' with other entities—branches, snakes, sticks—that were going in other directions and had other ends and goals."[64] She is profoundly aged by the work, her body marked and scarred and hurt. Her twin brother, on the other hand, travels in an air-conditioned, first-class compartment, reading his newspaper (and because, as we know, he is not moving in relation to the train he feels himself to be at rest, stationary). His three hours of uninterrupted, uneventful travel pass by almost without his noticing and he has aged only by those three hours, an entirely negligible degree. He will remember next to nothing.

Latour has begun the essay with an account of watching a windsurfer on a lake while he is on a break from a conference in Switzerland, full of psychologist and phenomenologists, most of whom assert the priority of lived time over the "objective" time they suppose that scientists produce. (Readers will remember this from my discussion of objective and lived time in Chapter 1.) Latour notes that he could, given some simple instruments, measure the speed of the windsurfer and calculate the ratio of distance over time. But he wishes to introduce another dimension between lived and objective time. In fact, he goes on to propose five dimensions by which times and spaces are generated. So in this moment of gazing at the windsurfer, he slyly converts the ratio of distance over time into another, which he calls transportation over transformation; this, he says, is "prior to the fabrication of times." Because of each traveler's differing relation to transportation and transformation (the twin on the train and the twin cutting through the jungle), "the production of times and spaces . . . will be entirely different." Relativity here is based on the relative experiences of entity-less smoothness, which Latour calls transportation, and entity-full intermingling and labor, which he calls transformation. In the case of the twin who journeys on foot, the one who makes the trail, movement and displacement will be associated with "modification, aging, history, transformation, metamorphosis."[65] She will not see any difference between "space, time, and aging."[66] The twin on the train will be able to differentiate between what seem to him to be "apparently different phenomena." He will perceive his own experience of displacement as moving through space in time, and, quite separately, the phenomena of "aging, living, suffering, participating in events."[67]

Latour stresses that the relation between time and space as a kind of "immutable" frame and space and time as caught up in "entities, beings, or events" is not a given but is specific to historically situated means of transportation and to the traveler. We can begin to see here the contours of Kentridge's colonial Africa and its railroads, its subjection to Meridian time (as we shall discuss soon), or the smooth and good roads built by Israel over the heads of the Palestinians in the Occupied West Bank, to which they have no access, in order to provide smooth passage back and forth for the Jewish settlers. For Latour, time and space are neither "the Newtonian sensoria" or the Kantian *a priori.*

Instead, space and time are "consequences of the way bodies relate to one another."[68]

Latour extends his thinking as follows. The twin on the train experiences what used to be places along the train's route, even places where he once enjoyed disembarking and visiting, as instants along the train's route, as the train speeds by them. The people who live in those places, by the same token, used to be able to get on the train, but now the train forms obstacles to their ability to move around. They have to take detours, they have to wait in order to be able to cross, and so forth. Latour suggests, then, that the train twin is an "intermediary" who does not have to take others into account and has no history or interruption. The twin on foot, however, is among the "mediators" who are "defining paths and fates on their own terms."[69] The use of the term fate is interesting here, as it points to Kentridge's work to put fate in his own terms, presumably an oxymoronic task, but one that for both Kentridge and Latour is possible. Fate, the pre-determined series of incidents that bring us to our end, does not define the work of the mediator, but becomes one of the mediator's mediums of transformation.[70]

On his train journey, the twin who is the intermediator, availing himself of transportation, depends on "the complete *obedience* of the places that are traversed," and upon the smoothly running clock time of the train system. But, says Latour, what happens if there is a protest among those through whose places and lives the train cleanly cuts? What if they begin "sitting on the tracks or even putting logs on the rails and setting them on fire?" Well, then, the intermediator is stopped in his tracks. And he begins to age. His speed is hampered by his forced relation to other entities. This town, having been reduced to an instant of time as the train moves through it, can, "because of this very revolt, become a place, a site, what we could call an event-producing *topos*." The revolt makes the train passenger feel time. He will have to choose alternate routes, he will have to encounter protestors shouting at him, protestors who are "making history," creating a "memorable spot to be reckoned with." And what, Latour asks, would happen if at every point along the trip, including on the roads that the buses are using to transport the frustrated train travelers, the revolt is happening, blockades are set up? Well, the smooth path for the train, for transportation, will have disappeared, back into the jungle. And this means that "it would be impossible for anyone to go straight through without being deeply and lastingly modified."[71] These are "revolts where space and time are decided on the spot,"[72] the initiation of history. And when a place becomes a topos, when it has stopped the smooth, uneventful, speeding train dependent upon the total obedience of the landscape it is traversing (the green expanse in Schneider's YouTube video, the smooth gray path standing in as tracks), it also, says Latour, becomes a kairos, by which he means a place of inception, a topos-kairos, an "event-producing spot."[73] He says, "Deeper than time and space there is another question about who or what counts. Which actants can interrupt, modify, interfere, interest which others, thus producing as many *topoi-kairoi*?"[74] Both Schneider and Kentridge produce topoi-kairoi. Schneider when he derails the YouTube video, when he fights the smooth time of the theatre, when he pulls back against his fate, his death from his fall, introduces modifications and transformations, entities, science lessons, novels, auditions, philosophers, doubles, electricity, LED, music, buzzers. It all starts when he derails that train. Penultimate time can here be imagined as a topoi-kairoi of creative innovation. In Kentridge, it is the studio that is topos-kairos at the heart of his practice, and at the heart of *Refuse the Hour*. The studio is where the mediator works.

And so, in *Refuse the Hour*, the revolts where the train, so to speak, is stopped, where time is felt and changed, are tracked through the institutions designed to keep time on its track, and to keep time the same for everyone, as established by zero-degree longitude, the Prime Meridian at the Royal Observatory of Greenwich. From the Greenwich Meridian, time was synchronized increasingly across the world, down the west coast of Africa, to Vietnam, to the Andes, creating, as Gallison says, "A planetary machine that would bring the world under one ticking clock."[75] Kentridge calls Greenwich "the colonial observatory." By means of cables, along rail lines, under ocean, "time was taken from the master clocks of London and Paris and sent to the colonies .... These strings of cables, these bird's nests of copper, turned the world into a giant switchboard for commerce and control ... local suns were shifted further and further from local zeniths."[76]

The colonial revolt appears onstage, mixed, found, in the studio. Ann Masina sings the aria of colonial revolt, "Give Us Back Our Sun," singing away the control by Europe. She sings furious and consolidated, planted center stage, square, and when she is done she flips her massive body away from us, so that her long skirt swings out in a gesture of total repudiation. On the studio screen there are overlays of old colonial maps, montage, with lettered pieces of paper, a litany of revolts: the 1898 Sierra Leone Hut Tax War, the Sudan Mahdi Revolts (1900–1904), the Chilembwe Revolt of 1915, and the Herero Revolt.

Kentridge and Gallison combine the colonial revolts with a bundle of attacks in 1894 by French anarchists, and in particular the attack by Martial Bourdin, meant to blow up the Greenwich Observatory. The bomb he built exploded as he climbed up to the observatory and killed him. Kentridge made five films (in the studio), using paper sets and live bodies. They bring together five locations to make their linked apparatus visible at the same time, even though they are not simultaneous in history, so that all together they create the explosion. Latour, speaking of the photographer Etienne-Jules Marey, a contemporary of Einstein who famously produced the images of the stages of doves taking flight, says Marey invented the "anti-movie camera! Something that would turn movement into a succession of images synoptically and not successively visible."[77] It is a way to see that is not successive, to see the linked apparatuses of time production and revolt.

The images in the films (the cover of this book is an example) are in multiple scales, painted paper mixed with real objects, white drawings on black backgrounds of clocks, equations, disoriented, overdetermined, the surrealist cartoon influence of Méliès, steampunk, *Eraserhead*, and maybe a little Charlie Chaplin in a sometimes humor-edged manic engagement with the machine. These are:

1. The Clock Room (Greenwich) 1894. This room is filled with paper versions of all kinds of clocks, spools, gauges, string unraveling, a white man in a white outfit working in the aggregation. It explodes, goes to Kentridge's charcoal oblivion. In the aftermath everything is askew, slanted, sheets of paper covering the floor, perhaps clocking notebooks destroyed. Dada is now present. She dances in the wreckage.
2. The Engine Room or Colonial War Office 1919. This is meant as the "machine room of the Empire" but also as the bellows, the source of air, the pneumatics controlling time, but also inflating a white paper globe in which a black man is encased. A white man works a huge object lung that looks like an accordion. Dada dances in the wreckage here, too, in white fabric that makes her look like a dove alighting.
3. Club Autonomie Dakar 1916. This is the place where the anarchist bomb was built, moved to Dakar, Senegal, to stand for revolt against colonial rule. There is the bomb, made as logs strapped together, with an old clock on top, wired, the end of the logs alight, as if the fuse is already lit.
4. The Map Room, the London Telegraph Office 1902. In this room, the huge white global encasement inflating a black man is being drawn upon by a white man (is it possible he is supposed to look a little like Einstein with his head of wild white hair and mustache?). He is maybe drawing lines of longitude. Mapping the world. Dada, as in all the situations, dances. Later, just before the procession to the black hole, the man's balloon is restored to a clear white surface, unmapped, colonialism undone, and he and Dada dance like children, laughing and bouncing off each other.
5. The Royal Observatory 1905. This space is filled with a large telescope and little models of the universe, stars, planets, suns bobbing on wires. Clocks are coordinated to these bodies.

The dynamite shatters it all. No more smooth travel.

In *Six Drawing Lessons*, Kentridge talks about trying to stage the Herero genocide. He is here so clearly Latour's mediator, laboring among and with many entities. He writes:

> There is a coming together of that which is in the studio: the dividers, the Erector set . . . and that which arrives from outside: Chilembwe's letter . . . Mozart's music . . . the chronology of the invention of Africa. Calling history into the studio . . . . The studio becomes thick with geography and time, upwards and backwards.[78]

Mediation, transformation. A revolt "where time and space are decided on the spot."[79] In the penultimate time before the procession to the black hole.

Kentridge's traveling twin example is full of entities, times, metaphors, losses. When the twins separate, and one travels into space, they wave goodbye, sending and receiving, as the traveling brother's wave gets "larger and slower" until it seems not to move at all, so that "the unmoving upraised hand pushes the brother away." But the traveling twin perceives the wave of his brother, standing "on the shore," as more and more insistent, frantic, "until the farewell gesture becomes a vibrating fist."[80] A fist is raised against the palm, pushing the other away. They seem to be separated irrevocably. The traveling brother, a combination of Perseus, Odysseus, of the ancient travelers, who go and then return, has so much to go through, so many adventures that he feels he has been gone for a lifetime rather than twenty years. (Of course, in the classic thought experiment, time has sped up for the traveler, even though he does not feel it since he is not in motion relative to the spaceship he is in.) But for the brother at home, waiting, for sixty or two hundred years, the years pass, one after another; everything is always the same. He ages to a point beyond memory. When the traveler returns, with his travel adventures, with his twenty years' worth of a lifetime of excitement, presumably bearing many scars, his brother does not recognize him. The laws of physics have been encumbered with human separation, starting with a gesture of farewell frozen into anger and renunciation. The ancient travelers return. They were sent out and they made their way back, going backward. They are still young. The ones who did not go, who did not send themselves out, who excepted themselves from the tasks involving mediating with other entities in a strangely accelerated and de-accelerated time, who only marked time by its incessant tick tick tick like the metronome, an unbearable sameness of time, are diminished into an unrecognizing cataracted age.

Kentridge's example is the reverse of Latour's; the traveler who moves speedily through space is not an intermediary but a mediator. Journeys can go backward and forward, journeys outside the tick tick tick of time, uncoordinated to the meridian, are hazardous, but, dare I say, they keep us alive. In the studio, they are in a penultimate time, alive against the pressure of death, in the creativity that comes before. Outside the studio, the time of the in-sync, the time of the metronome. Latour says of trying to find absolute zero, one of "the most daring scientific enterprises of this century," that it was driven by the search for constants. For Latour, there is an obsession with "immutable mobiles,"[81] things that can move without deformation, staying outside the studio "to make sure that every stage will be regulated according to schedule . . . that constancy will always be maintained in spite of the turmoil of history and world wars; that capitalization will go on forever without losses or spending."[82] In Kentridge's traveling twins example, the choice of constancy and the persistent clocking of time into death is the refusal of penultimate time. Kentridge finishes his monologue in this beautiful way. "There is the black hole, death, that swallows everything, Perseus, Danae, the grandfather, the eight-year old on a train journey with his father." But, speculate some physicists, entropy, the disintegration of energy, prevents everything from being absorbed. Perhaps at the edge, at the event horizon, entropy has left in its wake "vibrating strings, twists, knots, cat's cradles of information."[83]

Revealing the secret of that exhilarating advance of the musicians, actors, and singers straight downstage to the audience, even as the shadow procession on the film behind moves laterally toward the black hole, Kentridge says:

> The bank at the edge of the River Styx, where Charon deposits those headed for the black darkness of Hades. He keeps in his boat the attributes they have shed. A suitcase of teeth and glasses, thoughts, stories, an old stone discus. Held in trust on account, waiting for them to be decoded, to be reconstructed, to be made new.[84]

This is what we in the theatre do. At the edge, just before, but not in. Where we can make time creative, ours. We make the attributes new, bring them back, return them, talk backward, send and receive. Each performance a new beginning, an inception, at this edge, before death. In the studio.

Before Your Very Eyes

"You know why we're here: we're here to live and die." At the very beginning of this chapter, I quoted this directive to the child performers in Gob Squad's *Before Your Very Eyes*. You may recall that in the performance a group of children enacts successive stages of their lives and, ultimately, their deaths. Although on the surface of things we might take this performance as representing precisely the inexorable tick tock to our deaths, that time in which we are and in which we have no say, *Before Your Very Eyes* is really about dancing back from death, or dancing before death, in penultimate time, just as in Schneider and Kentridge.

The show is a highly complex negotiation with the future, with our common and widespread stakes in children as the future, as the figure by which chronological time unfolds into a future that will be lived beyond our own deaths and to which we attach our hopes for a "better world." It is a complex negotiation with, and ultimately a parody of, the idea that we are only here to live and die. In a sense, *Before Your Very Eyes* is double sided. It convincingly and often movingly represents the sequence of life's disappointments that perhaps many in the audience have experienced, or, if they are young, fear that they will experience. For instance, the first time I saw it, I was with a twenty-something former student, a brilliant, talented, and underpaid art director's assistant for a daytime TV show. A precarious worker. He wept at what he felt was the show's confirmation that he would never succeed in getting what he dreamed of. I wept at poignant moments about loss and death. There are many times when the silliness of the show turns to genuine pathos. In this way, it seems to genuinely show the future of the children as already determined by disappointment and then death. They are only here to die. The show seems invested in the future, but elegiac about what the future means now, in neoliberal times.

But it is my argument that the show is really doing something very different. Ultimately, it is a joyful parody of the idea of the future. It withdraws from the future, withdraws children as figures of the future, and is in fact the representation of a chronogenetic, operational time. In my description of the show it will be clear that it would be very possible to interpret it as embedding the children in a sequence of life's disappointments, children as the future foreclosed but unable to stop moving into it. But I propose a different kind of analysis that proposes instead its life affirming, joyful, making of time. In so doing, I draw on a fascinating essay by Walter Benjamin on children in the communist theatre, and their position in relation to the future. This helps to move us closer to replacing the future with something else, relieving children of their futural obligations, and as such their obligations to processional history—as I believe Gob Squad does with the children in this play. They are put in relation to what is on the surface established as the certainty of death, the destination that determines all of life. But they are provided by Gob Squad with the means for the temporal innovation that, as in Schneider and Kentridge, happens in the creation before it.

I saw the show twice at the Public Theatre in New York City in spring 2016. I should note that the quotations from the performance are all from notes I made while watching it, with some help from a promotional video. We settle down into our seats in front of a large clear box on stage. Inside, children aged 8–14 are variously playing, being "childlike," horsing around. The walls are mirrored inside, so that the children do not see the audience watching them. As with Gob Squad's work in general, there is a lot of technology. There are screens inside the box and screens outside, startling transitions from virtual to embodied and back again. The children seem, or are acting, unaware of our presence, although they must be able to hear us. More spectators fill the theatre and the theatre quiets. The voice, a soothing, even-toned female voice, says, apparently to the children, "I just want to let you know that they're all here now and they're sitting watching you." On a background screen this flashes in red, in marquee lettering done in bulbs, side show or vaudeville style, playful.

> Ladies and Gentlemen!
> Gob Squad Proudly Presents
> Real life children!
> In a rare and magnificent opportunity
> To witness seven lives lived in fast forward . . .
> Before Your Very Eyes!

The show begins. A child with a microphone says "We've been thinking a lot about death lately. No matter how things work out we'll be dying. We want to face this without fear. We are all speeding toward death." The voice responds, "Okay, but I don't think that's what they're all here for. You're all so young. They want to see you carefree."

The box darkens, or empties, I cannot remember. There is some kind of magician's trick, and then the children appear, one by one, on a screen outside the box. Each child is dancing, dancing on the streets of New York City. (The piece was also performed in Belgium and Berlin in 2011. For each performance, Gob Squad worked over a period of two years with children from that specific city.) So here New Yorkers see each child in familiar cityscapes, dancing with happiness and abandon, hair flying, arms and legs in every direction, smiles and bright eyes. There is music, as there will be throughout the show, popular songs which mark the life "phases" the children will enact. Right now, they are dancing to Queen's "Don't Stop Me Now"—driving beat, exhilarating rhythm: "On a rocket ship on my way to Mars / I'm burning through the sky / 200 degrees that's why they call me Mr. Fahrenheit traveling at the speed of light / no stopping / no stopping."

Each dance is terminated with the ejection of the child from the film image back into embodied reality inside the box. The velocity with which each is thrown makes them have to adjust so as not to fall when their bodies jolt to a stop inside the enclosure. Acceleration into the future is established as a theme. The children accrue inside the box, one by one, as outside on the screen another is still dancing. The last child barely makes it into box before the music cuts out, abruptly. She keeps dancing silently as the others watch her. She is making sort of a spectacle of herself. She does not have the behavioral code quite right. The voice says repeatedly, "You can stop now," which she finally does. (In New York there were two teams of children performing at alternate performances. If performed with another group of children, this last child could have been a boy.) Inside the box there is only this group. It strikes me that inside this box there are no other people, no politics, no news, no family. At first, in this most obvious face of the show, this seems to be an adjustment from a world to solipsism, the joyful body suddenly dulled, or many of its multitudinous parts suddenly eliminated. After all, they are here to represent life's passage toward death. The voice says, "Can we hear her heart now please?" Another child brings a mic and kneels down in front of the last dancing child and places the mic in front of her heart. We hear the heartbeat. The voice says, "That's a beginning. The sound of a small heart beating." It is the evocation of the fetus in utero, when the doctor first puts the stethoscope to the stomach and one hears the heart for the first time, the promise of the future. There is a pause, and when the voice resumes this futural moment it is turned toward death. The voice says, "You do know why you're here don't you? You're here to live and then die. I don't think anyone can sit here for decades so let's get on with it. Come on. Grow up." The children leap into action, joke, playact growing, pretending to stretch each other's bodies and their own into exaggerated lengths.

The voice then asks them to turn on the TV inside the box. The voice says she wants to show them when they and the voice first met. The screen on audience left, outside the box, shows each child, one by one, as their two-years younger self. Each child says what they imagine their future will be. These are bright pictures, the kinds of good fantasies that children manage to have about how their lives will turn out. Many children in this markedly privileged and relatively homogenous set (a set of New York City children whose parents have been able to support this activity over two years) dream of those kinds of successes that will help them save humanity, save a species. One child in a performance I saw, the 11-year-old version of his now 13-year-old on-stage self, gave a rendering of "To Be or Not to Be" that embarrasses his 13-year-old counterpart. The voice says to him, "See, Elijah, you have grown." Elijah responds, embarrassed perhaps to have had others glimpse this innocent and naïve smaller version, "Fuck you." Thus, we have in place a contrast, which will be revisited later in the show, between an "original" or more innocent or less jaded beginning of each of these children who will now enact, guided by the voice, their successive arrivals at age 20, age 40, age 45, and death. Each of these "stages" is represented in the clichés of white, middle-class European/North American privilege (even though some of the kids were not white, white privilege was the dominate paradigm). All children, regardless of age, performed each age, so that we saw 8-year-olds as well as 14-year-olds performing 45. Each age was done in the height of theatrical shtick—terrible wigs, too big clothes, hastily and obviously done make-up, like Halloween dress-up. Each age was accompanied by a pop song, the lyrics and the tone of which underscored the particular affect of the age. The voice interrogates them at each age, asking, for instance, "So what can you do now you're 21?" At 21 they sing the Pierces' song "Boring": "Nothing thrills us anymore / No one kills us anymore / Life is such a chore / When it's boring."

Sometimes the voice issues instructions, and the children follow them. In the age 40 scene they have donned their middle-aged selves—big cheap mustache, bad wigs, fat bellies sticking out of undersized T-shirts, too much lipstick. The voice directs them through the scene: "It's your fortieth birthday. You never thought this day would come. It's your birthday, but it's a disaster. Offer the sushi again. Talk about wine," and so on.

At age 45 they are defeated. When the voice asks a girl what she can do now, the girl answers, flat and listless, "I can put fruit in a bowl just for show. I can get Botox to try to look younger. I can worry about getting old and crazy surrounded by cats. I can say I'm about to have a burn-out." Each stage of life is equally affect-less. There is no interruption in the ennui resulting from the linear progression of these relatively privileged lives. There is no disease, no harm, no poverty, no disaster, no early or unexpected deaths. Just this lead up to death in old age. In one of the truly poignant moments, the children come down and spread out laterally against the downstage wall of the house, as if gazing outside. They karaoke a 2008 song called "Wolves" by the band Phosphorescent. It is a son singing to a mother, talking about wolves/wildness tearing up the house, inhabiting the heart. It seems like a mid-life crisis song, about a repressed interior life full of the desire for beauty and wildness.

As the song finishes, the voice says, "Good. You've lived half a life now. What's wrong? Have things not turned out as you expected? What did you expect?" There is a repetition of the scene with the young "original" version of the self, but this time the younger children, rather than describing their hopes for their future selves, speak to the older version to directly ask if they attained the early dreams. The worlds the children were ejected from into the box, which clearly is also a spatial metaphor for "boxed in," those worlds with politics, geopolitics, imagination, art, skills, singular capacities and relations, happiness, decisions, intrude briefly, again. Some of the dreams are positive, left-trending, but not all. One of the young versions dreams, extensively, almost rabidly, a hypercapitalist fantasy of how rich he will be, rich and famous, a star-studded career, many houses. Somehow his older self leaves the box and, as the others karaoke Edith Piaf's "No Regrets," he walks up to touch the box from the outside as on the screen the younger version's hypercapitalist fantasy goes on in silence. He gets out, but clearly there is nothing there, on the outside of the box he lives in, so he goes back in. The Piaf song scratches to a sudden halt. He is now subjected to the voice's instructions almost as if being punished. The voice commands him: "Now say: What was an exciting mysterious future is now starting to fade. You realize you are not special as the world forgets you . . . as you learn no one is watching you and never was." Boy: "Who's in charge? What's coming next." Voice: "Well, it's obvious." All children: "Hurray! Let's do the dying scene now!" The fantasy of the future is unattainable. It has receded. Death takes its place and is the only destination.

In so many ways, this face of the show could be an illustration of Lauren Berlant's intensely pessimistic, brilliant, and much relied upon book, *Cruel Optimism*, mentioned in Chapter 1, and its central term: the "impasse." The impasse is a state of being and surviving which entails a kind of "dogpaddling"[85] in a situation in which people find themselves adrift and in a constant process of adaptation without any registers of certainty or knowledge about what it is, exactly, that is happening. In particular, as a neoliberal form of time, it is a "survival time" that, as in the opening song of *Before Your Very Eyes* ("no stopping, no stopping"), is a "time of struggling, drowning, holding onto the ledge, treading water—the time of no-stopping." The children are reproducing "what we should not call the good life, but the 'bad life'—that is, a life dedicated to moving toward the good life's normative/utopian zone but actually stuck in what we might call survival time."[86]

At first sight, it seems overwhelming and not a little coercive that the representation of these children's lives and their original hopes for themselves in the future diminish to a form of survival time. To me it seemed that, since Gob Squad worked directly with the children over two years, the hopes and disappointments and images of themselves in the future were verbatim, gleaned directly from the children themselves. This begs the question of how it was possible that the children were somehow already capable of imagining the totality of their aspirational failure, the dogpaddling in survival time, until death. Is the impasse that Berlant describes so profound that it has already saturated children above the age of 8? Or, is it really about the voice? Is the voice the representative of the inscription of the desolation of being carried along by time as a spectator of one's own life, helpless? The time "*in* which we are," rather than the time "*that* we are," to return to Agamben? Is Gob Squad, with the voice, doing something like what Ant Hampton does in *The Extra People*? In that piece, a computerized voice directs workers in huge warehouses where the only paths of motion are up and down and cross-wise huge depopulated aisles, their only companion throughout the day a simulated intelligence ordering pick-ups of products to be shipped. At the end of the performance, we are shown videos of how this works in a real warehouse like an Amazon warehouse. Disobedience is unimaginable. At a particularly poignant moment in *Before Your Very Eyes*, the boy who had briefly left the box and returned, when told that his "mysterious exciting future is beginning to fade," says to the voice, "who's to say we wouldn't have done it all completely differently? Do you believe in fate?" The voice replies: "I can't tell you. I'm just a voice." The voice is person-less, incapable of answering the first and only philosophical question raised by the children, on the question of determination and choice, fate and decision or chance. The voice is pre-programmed, a computer not designed for a non-obedient and unexpected question, a child's question voicing something like Kentridge's protest against fate: "This is intolerable!"

But something changes before the end. The children, having gotten to the dying scene, are dancing. They are in their comic get-ups as the very old, age lines painted theatrically on their faces. It is a desperate dance, once again to "Don't Stop Me Now." It is grim, aerobic, in formation, facing downstage. The children are aping aging bodies trying to keep up. But they keep dropping "dead," literally, on the spot, until the floor is littered with their bodies in various hilariously caricatured poses of types of death, the kind children would imagine and playact. Finally, only one is left. Not only is she about to die, she is suffering from "being left behind." It is a grim scene. The voice prods her with questions. She says she wishes she had traveled more. The voice asks her, "What happens when you die?" She says, "They put you in a box and carry you away." There is no question of immortality. When she slumps and dies, the children, one by one, begin to back through the door that opens in the box's wall stage right.

But, here is the change. Once through, in a sequence that mirrors the opening of the performance, each child is passed through the wall onto the film screen, where they then, again, dance, wildly, with abandon, with joy. Onscreen, the other people in the world of the particular cityscape are moving backward as if the child dances back, dances away from death. Time, as in Kentridge, is reversing. Joy is dancing itself back into the time before death. We feel joy. The children are not the future. There is no longer the future. They are signaling back from this operational time where clearly time is changing. This dancing back flows back onto everything we have already seen and changes it. To me, what I have already seen becomes full as an operational time.

In 1929 Walter Benjamin wrote a short text on children playing called "Program for a Proletarian Children's Theater." As Nicholas Ridout details, it was the result of his agreement to work with Asja Lacis, a director from Latvia, to develop a children's theatre program for the German Communist Party. Benjamin's text hinges precisely on the problem of children as the future, or, to put it differently, the question of how to educate children who will become workers for and builders of the proletarian future. As Ridout puts it in the context of his study of amateur theatre, which abandons the models of professional labor by which theatre is yoked to capitalism, Benjamin saw the children as amateurs who should be neither "in the service of capitalist development or even in the teleology of the revolutionary project." According to Ridout, Benjamin is writing "anti-programmatic thought into the party program."[87] For the professional (including the Soviet professional), work "toward the construction of an ideal community is the dominant mode in which history might be experienced or enacted."[88] Benjamin's stakes in the redefinition of temporality, history, and the redefinition of historical progress (processionism) means that he sees the children and what they do in their theatrical training as, according to Ridout, "disruptions in historical time . . . at the level of the everyday."[89]

More broadly, for Benjamin, children are beginnings, the start of the capacity for revolutionary transformation. Children's mimetic gestural play with things is continuously releasing new forms of meaning—nothing sediments. This is a revolutionary model because it escapes "bourgeois socialization . . . . Parroting back the 'correct' answer, looking without touching, solving problems 'in the head,'"[90] in Susan Buck-Morss's reading. In her comments on Benjamin and children, she notes that both Piaget and Benjamin agreed that the child's forms of understanding were eradicated in adulthood. But while Piaget was content with this since he was more interested in the adult mind, Benjamin saw this inattention to the child as grounded in "that assumption of history-as-progress which Benjamin considered the trademark of bourgeois false consciousness."[91] For him, children were instead "representatives of Paradise," and "revolutions appeared, not as the culmination of world history, but as a fresh start."[92] For me, the word "paradise" temporally transcends the future, and therefore in Benjamin's ingenious language, children are positioned in a time independent of determination by the temporal mechanisms of processional history, a secular paradise of the everyday, of the fresh starts and beginnings of children's play, in its lack of coherent resolution.

To return specifically to the theatre, in his Program Benjamin differentiates between bourgeois education, in which children are led toward a set of ideas that they will come to know and master, and proletarian education. In the proletarian education, children will instead be inside all of life, and "its entire life [will] be engaged."[93] This is why the theatre is ideal site for their education. The children, he says, will work as a collective which "radiates not just the most powerful energies, but also the most relevant ones." Unlike the bourgeois spectator, who looks for the "moral personality" of the director manifest in the children who he is leading toward a future set of actions, the proletarian audience practices an "unsentimental love"[94] in close observation of the children to see how "every childhood action and gesture becomes a signal from another world, in which the child lives and commands."[95] All the adult leader does, says Benjamin, is to help the children apply their imagination to materials. The goal is not "the 'eternity' of the products, but the moment of the gesture."[96]

Benjamin finishes the essay by suggesting a temporal politics in which children are central and which will "annihilate the pseudo-revolutionary gestures of the recent theatre of the bourgeoisie," based as it is on what we would probably today call hitting people over the head, but what Benjamin calls the "propaganda of ideas." Instead, he says, "What is truly revolutionary is the *secret signal* of what is to come that speaks from the gesture of the child."[97] Benjamin, it would seem, deliberately replaces "the future" with the phrase "to-come" in an explicit repudiation of propagandistic plans for the future in both bourgeoisie, left bourgeoisie, and revolutionary political and theatre practice. The children's signal is secret, as if to protect it from the heavy hand of the processional history that would inevitably spin it out into something coming from a past and belonging to the future. The word "to-come," as rendered in this translation of Benjamin, stands in for what, paradoxically, does not move on, but initiates, begins.

In a short promotional video from the Public Theatre, Simon Will and Bastian Trost of Gob Squad talk about how happy they were that they were able to "transport" to the children Gob Squad's method of performing themselves and the freedom that it gives. He says that being in the box where they could not see the audience looking at them put them "at ease," "amongst themselves," with "a freedom to react to each other," "to be like a team, like a little collective."[98] I think we can understand the Gob Squad children as a collective, playing, in the fulfillment of their childhood. They are playing, faking, signaling from a world they are commanding. They act out death in a child's way, in a theatrical way, far from death. They do not succumb to death, their future. Observing them, in their box, observing their gestures, their putting on of the deadly enervation of bourgeoisie life in obedience to chronology, acting out their obedience to the voice of that enervation, we should try to see the signals of the to-come which is not the future, but something that is a break in processional history altogether.

The show makes the theatre itself an operational time. It spoofs theatre's assumed inextricability from death. In its spoofing, in its play, in its stupid costumes, it sends up forms of dependable narrativity that the theatre is supposed to provide, birth to death and the events in between, even as it embeds the performance in multiple symptoms of contemporary neoliberal life, themselves ultimately transcended. And the fact is that the children have already exceeded these predicted lives. They are children who are actors, who are creating, who come over and over to a space where they make again, begin something again. In this operational time the youth and energy and creativity of the children is made and sustained in a way that cannot be consumed by the frame the piece purports to create at the beginning, the representation of time as moving toward death. The children's out-of-sync-ness with the chronological representation of time is what allows them to "take hold" of time, to replace the future with the to-come, even as they are still (biologically, realistically, and through the set up for the theatrical representation) in chronological time. Inside their mirrored box, they are reminded of their youth all the time. In the difference of their youth from the narrative they are playing at, they are in charge of making both the representation and its out-of-sync interior—they have hold of time. It is "the time *that* they are." I think this is the secret of what Gob Squad gives these children, this "collective."

In *Before Your Very Eyes* the end, as death, is staged, but as something near to a mischievous joke, with hokey poses performed by children in dress-up clothes and age make-up. It is so clear that there is no end here; there is no attempt to represent an end in theatre's best traditions of death scenes. If the show begins with the injunction that the children are here to live and die, that each one of us is going to die, its sleight of hand is to pull that death out from under us, even as it pulls on our hearts, a little. We are not allowed death. We are not just here to live and die after all. In essence, death is skipped over. The theatre has provided a time, a penultimate time, a time of creation, rather than an end, a death, anything finite. The very idea of chronology is meaningless as those children dance away from death.

On the Concept of the Face: Regarding the Son of God

Another group of children onstage provides us with another model of operational time in penultimate time that is overtly, it seems to me, about the initiation of history. These children appear in Romeo Castellucci's *On the Concept of the Face: Regarding the Son of God*. The performance, the main subject of which Castellucci says is time,[99] documents in excruciating detail, and almost without language, the moment-to-moment progression of an old man's uncontainable diarrhea, uncontainable shit, as he, presumably, moves toward death. It is a pristinely white, decidedly upper-class space, with couch and coffee table, glass desk, and bed and bedside table, along with some tactfully placed medical paraphernalia. The stage image is dominated by a huge portrait at the back, Antonello de Messina's 1465 painting *Cristo benedicente*. The image is cropped so that only the face of Jesus, chin to forehead, looks out at us, the face filling upstage center from the deck of the stage up into the fly space. It is an impassive face, deep but unreadable. The expression on the face is defeated, or compassionate, forlorn or assessing, but ultimately distant. The old man is dressed in a white bathrobe and is diapered, sitting on the white sofa as the performance begins. His son, dressed in the costume of a successful business man with a pure white shirt, cleans him over and over, with increasingly less effectiveness as the shit comes faster than he can clean. His father sobs; he whimpers; he is clearly undone by the humiliation, more so as it gets worse and worse; he says he is sorry over and over. The son, devoted, ceaselessly patient as his departure for work is continually delayed by the next episode of leaking shit, wipes his father's body (with white towels), wipes his father's hands after the father has already rested his forehead in them, getting his forehead covered in shit, moving him from the couch to his walker, perhaps to get him to the bathroom only to have the old man once again, and worse than ever, expel what is now only brown liquid all over the floor. The white floor begins to run in places with the brown liquid; the son cannot keep up. His major gesture is to console, but at times he is overtaken by a despair, expressed gesturally, minimally, and at a very few points he gives in to anger, an explosion of repetitive words. Eventually, the son takes the old man to the bed and leaves. The old man seems to make a decision. He takes a gallon-sized clear plastic container from the bedside table. It appears to be filled with shit. Standing, he faces the bed and pours the shit all over it. Then he sits down, facing out to the audience, elbows on knees, head bowed. The face watches the progression of chronological time that disintegrates bodies and minds, leading toward inevitable death. Helpless, we watch God, regarding us from somewhere unreachable, a changeless time, a cosmic surveillance. The face regards the old man (us too, perhaps) as life lived only as sequence, moment to moment, ground down to nothing more than that. The old father is alone, afraid, infant, helpless, having achieved nothing.

But then the first child, about 10 or 11 years old, walks on with a backpack. He stands to look at the face. The theatre has filled with a sound. It is as if it comes from behind the face; it sounds like a great metal room, a room that has walls but is as big as eternity, as if something were hitting the walls and bouncing forever, as if a wild game of cosmic handball was underway. There is also a high whining whistle sound like a firework going up, except this sound has no directionality. The total sound is curved, so curiously both closed and infinite. It is like the voice of this face. Nothing of the elements of the sound changes, only a kind of randomness of that sound of balls hitting, the whistle—as if it has been there forever and will be there forever, unassailable. The eyes in the face of God are compassionate, but the sound of God is an iron determination of the infinite that does not correspond the experience of the suffering old man on stage. It is compassion without intervention.

But the child empties his backpack onto the floor. It is filled with grenades. He chooses one, straightens up, pulls the pin, and hurls it at the face; it is a long distance and high up; it takes strength and a good arm, physical investment. The grenade hits the face and bounces off. When it hits there is a sound like a grenade exploded, magnified exponentially, and contained in that sound, an impression of mortar, foundation, crumbling. But as the boy waits, watching, breathing a little heavily, the sound of the effect of the grenade fades and the sound behind the face, the voice of the face returns, as if only mildly and temporarily perturbed, a slight and brief deformation. When he hears this, he picks up another grenade and throws it. The same thing happens as he again waits. After this, he begins to throw grenades faster and faster, with only the interval of pulling the pin in between. At the same time, other children begin to enter. They too have backpacks full of grenades that they empty on the floor, and that they also begin throwing at the face. They enter calmly, with purpose, as if somehow summoned, but also as if they had been ready. They walk on like laborers, like it is all in a day's work. They look at the face as if they have seen it, or others like it, before, and they know what to do with it. It is almost as if inside every sequence, an instant-by-instant process in which there seems to be no possibility for interruption, the children are waiting. The children await their function as chronogenesis, initiation. They arrive, so to speak, in penultimate time. They interrupt the time of the death of the old man and turn it to penultimate time. Insurrectionists. Stone throwers. They hurl their resilience and defiance into breaking the static oppression of chronological time. The children are not the future, but the inception, initiating a new possibility for what it means to live. The old man does not move, but he is brightly lit, drawing the eye in simultaneous focus. He is the marker of this time that is before death.

As the face comes under constant assault from more and more grenades, from about ten or eleven children in all, stretched across the downstage edge of the playing area, that voice of eternity is almost completely drowned out by the sound of destruction, masonry shattering and falling, chaos, rock exploding. It is a vertical sound, as if the debris deluge is starting from high up. The grenade throwing ends.

And, when it does, the old man gets up. Far from dying, he walks out, walks from downstage left to upstage right—free of this sequence, no longer carried along by its inexorable process, on his own two feet. There is no death, after all, in this play—which, at the beginning, gives every indication of a movement toward death. Castelluci absolutely breaks theatre's own temporal convention here. The man is retrieved from death. Death, simultaneous to the dismantling of the time toward death, is irrelevant.

The face begins to fade a little, and the voice of the face is there again, except that it is now hard to distinguish aurally from the barrage of the explosive work of the grenades. It seems to be in a state of extreme agitation, like highly heated atoms, as if the effect of the grenades, although fallen back on the floor, unexploded, stage props that are now past their use, has been to affect the very interior of this cosmic eternity. And the face is still there. Except for the sound, it is as if the grenades have not had an effect. A projection casts huge words across the face. They read "You are my shepherd" but fades to a word "not" that has been invisibly there. The phrase becomes "You are not my shepherd."

And then the door at the back of the theatre opens. We can perceive it only by door-shaped light. Then it closes. We wait. A dark blot, what looks like some kind of runny ink, appears on the face. At first it is almost impossible to tell what it is, or what is happening. But gradually it becomes apparent that what looked like spreading blots are actually portions of the photograph of the face being ripped down. We begin to see stage hands on scaffolding in the gaps created by the demolition. The sound is metallic, drone, squeaky chains, death gasping. It is as if the grenades—or, rather, what they stand for, the initiation by the children of their own history, their repudiation of power's time—have in fact dismantled the face from within, so that now it is only a huge blown-up photograph held up by a scaffold.

What is it that calls those children forth? Is it a "sudden act of consciousness" on the part of the old man, shown also by his decision to spread his liquid feces all over the bed? Not to shit helplessly, as before, out of control, but to choose to defile, to take defilement into his own hands. In another essay in which he discusses time, "The Critique of the Instant and the Continuum," Agamben says that the Western tradition is full, in its shadows, of suggestions for ways to critique the instant, to show the times of which Agamben speaks, the penultimate, the kairos. They are "bearers of a message" for us that "it is our task to verify." One of these traditions is the Gnostic. He writes, "The time of Gnosticism, therefore, is an incoherent and unhomogeneous time, whose truth is in the moment of abrupt interruption, when man, in a sudden act of consciousness, takes possession of his own condition of being resurrected."[100] Maybe the act of spreading the feces, followed by the entrance of the children, is truly a Gnostic moment. It would be hard to identify a theatre director more exploratory of the "folds and shadows of the Western cultural tradition" than Castellucci. I do not mean to suggest here that my reading of the scene was Castellucci's "intention." Rather, that attention to the moment allows us to experience that Gnostic message, another way to speak of initiation, to take possession of our own condition of being resurrected. Agamben continues, "In keeping with this experience of interrupted time, the Gnostic attitude is resolutely revolutionary; it refuses the past while valuing in it, through an exemplary sense of the present, precisely what was condemned as negative and expecting nothing from the future."[101] The negative of the past, the however many days and days of the old man's incontinence, is overturned in the taking hold of it in the old man pouring this shit himself. There is no expectation of the future at all in *Concept*—the children do not stand for the future here, there is no future, only this tearing down of that time image that guarantees subordination to a time *in which* we are, helpless, to a time always oriented to death as its remorseless destination.

Beginning in the mid- to late nineteenth century, childhood was invented as the place of the future, children as the bearers of the future in which we have faith and trust. In this piece, as in *Before Your Very Eyes,* children are assigned to a different use. The children are there to pull the future back into the incoherent and unresolved creativity of the to-come, which may be, after all, if Benjamin is right, the secret proclivity of childhood, a revolutionary activity that is not submitted to either capitalist or revolutionary chronology. The performances included in this chapter have all contributed, I hope, to theatre's capacity to be a medium for operational time, the time in which we take hold of time, the time we initiate. The frame for this has been penultimate time, and each has been not only an exploration of chronogenesis within this time, but the repudiation of the procession to death to which we are mostly trained to subject our lives. The replacement of the future with the concept of the to-come that the actions of the children in the last two performances create will now guide us into the final chapter, which takes up the question of the future directly.

4

The Volatility of Time in the Hold: Kairos and the To-Come

In this chapter, I write about four performances and the ending of one film. These are Forced Entertainment's *Real Magic*, Cassils's *Tiresias* and *Inextinguishable Fire*, Omar Rajeh's *Beytna*, and the end of the 1928 film *The Passion of Joan of Arc*, directed by Carl Theodor Dreyer (and with Antonin Artaud as Jean Massieu). I have gathered these together according to the glimpse each offers into a containment, a hold, an experiment in not moving on. I treat the hold as the formation and performance of kairological time, time stopped at the brink, time undefinable by the futural directionality of processional history. We are talking about almost indiscernible motion, near uncapturable interior motion, something that is making a new kind of time, inside what is held on the brink or in the hold. I will rely most centrally on a loose adaptation of Antonio Negri's construction of the ancient concept of kairos as what I see as an immobile generative force. We are not talking, necessarily, about an absence of physical motion, of stillness, although Cassils/Tiresias and St. Joan are both figures who are immobilized, one by ice and the other by ropes and fire. We are not talking about stillness itself as a kind of "resistance," although in some ways stillness can be an abdication from or a refusal of the logics and trajectories of productivity, self-production, self-valuation, or as David Bissell and Gillian Fuller say in their study of stillness, "outside a productivist relation to movement." But Negri's highly visual figure for kairos, of standing at the edge of time, leaning over the void, like the point of an arrow unleashed over it, is not an image of stillness, but an image of holding, in which the holding contains an active creation, an extension of imagination. Time is volatile in the hold. We are watching a "volatility of stillness."[1] We are replacing the future with the to-come, a process begun in the last chapter. If the new present is formulated in relation to the past, and the penultimate in relation to finitude and death, kairos, deliberately placed at the end of this book, is a dismantling of the future and its replacement.

Real Magic

There are the three of them onstage. Claire is in a chicken suit and head, garish yellow and comic, although we do not yet know it is she inside there. Jerry is in a ridiculously big suit with a long, curly brown banged wig like a 1980s rock and roller. Richard is in a T-shirt and his brief shorts. There are, on stage, a mic at center and plain plastic chair stage right of it. These sit on a rectangular cut of Astroturf, which also allows for more space on stage left of the mic. The turf effectively marks the performance area of this strange and strangely unidentifiable form of game show/quiz show or, as director Tim Etchells says in the program, something which is also an "absurdist cabaret mind-reading routine." The larger playing area is ringed by vertical strips of florescent tubes, hung from about twelve feet high to the floor. There are also a few sign-sized pieces of cardboard and other chicken suit pieces strewn around. That is it. The show, I will say at the outset, is very funny.

The outline, or procedure, for this show as game show is that one person is to think of a word, and the other person is given three chances to try to think of what it is. The person guessing sits in the chair and puts on a blindfold. The person thinking stands to his or her left, on the other side of the MC, who stands center at the mic. There are bursts of canned laughter and applause, which at first seem coordinated to what is happening on stage, but gradually become more and more arbitrary, off-time. The light tubes fizz and change color and otherwise punctuate the game. There is music, a kind of circus/entertainment music, possibly a canned version of the "Chicken Dance." And, sometimes, as the contestant tries to think of the answer, there is a loud clock, a tick tock tick tock sound that fills the space, that forces the imposition of accelerated game show time.

It is, of course, an impossible task—to pick the word someone is thinking out of all possible words. Absurd. In trademark Forced Entertainment style, the actors—Claire, Jerry, and Richard, keeping their real names—work on the line between realism as themselves and artifice. Or, rather, it is not so much a line as an integration, the integration of realism and artifice as themselves. They are never characters. The game is completely circular. Every time the MC asks the thinker if the guesser's word is correct, the thinker says no. Every time, the MC announces that the guesser has failed, and says "Let's swap." Every swap means that each of the three of them takes on another of the three roles in an endless rotation. At each swap they go upstage and put on something slightly different from the limited and recirculated collection of clothes. They interchange bits from their original outfits and part of or a whole chicken outfit is always in circulation on one of their bodies. Sometimes, during the swaps, a vamp from the "Chicken Dance" plays, with vamping a sound that is deliberately used to put an action on hold, indefinitely.

With every swap, Jerry, Richard, and Claire act as if they have never done any of it before, adopting slightly different affects, attitudes, and quirks each time they become, again, the guesser, the thinker, or the MC. Their extraordinary suppleness as performers, the fluidity and subtlety of the physical work, especially in their faces, produces a shifting and variation within each single person, each bounded body and allows them extraordinary variation within the same. It is never as if they are pretending to be different people. They always have the same names. Instead, it is as if they are willing themselves and each other to work with variations, to put pressure on variation, to start again with each variation. So, for about the first half of the show, each time there is a swap the MC asks each all over again: "What is your name?" ("Jerry," "Claire," "Richard") and "Have you ever met Jerry/Claire/Richard before?" ("No") and "Have you ever done anything like this before?" ("No"). These are always the answers, even though obviously they have met Jerry/Claire/Richard before, over and over, and have done this game before, over and over. But each time, at least until things start to fall apart, each of the three meets the situation with a pronounced ingenuousness as if each time were the first time. It is an infernal ongoingness of the same marked by a determined and increasingly desperate play consisting of highly constrained invention that goes nowhere except back into the same cycle.

The game works as follows. Each time someone becomes the thinker, he or she goes upstage and grabs one of the pieces of cardboard from the floor. It has a word written in magic marker on it. The thinker holds the sign facing the audience so that the audience can see what he or she is thinking. This position prevents the guesser from seeing it. For about the first twenty minutes of this seventy-minute show, each thinker, in turn, does some kind of silly business with the sign, with a wink wink, exaggerated "isn't this fun" attitude toward the audience. Each guesser, in turn, acts excited, nervous, highly anticipatory, giggly with all the fun of being on a game show for the first time. The MC asks each guesser at the outset, "Are you feeling good, confident, safe?" and the guesser responds with a self-deprecating, shy, shoulder shrug of delight, head nodding, smiling.

The thinker is only ever thinking of one of three words: sausage, caravan, or algebra. The guessers, in turn, only ever respond with one of three answers: money, electricity, or hole. Obviously, these do not and never will correlate. This does not prevent the guessers, however, from offering one of these three answers with a burst of happy inspiration as if it is the first time the word has been thought of and he or she is certain it will be correct. They are determined, each time, to produce the affect of happy anticipation, anticipation of the correct answer and the game moving forward, moving into a future in which they will be winners. But this will not, and cannot, happen.

The audience catches on before long. We understand that since these are the only three answers that will ever be given and since they will never correspond to the words on the signs held by the thinkers, there will never be a winner and therefore never an end to the game, never a movement forward into the future. (Of course, we also understand that winning such a game would be impossible under any circumstances. We cannot of course guess what someone else is thinking.) It is a case of discrepant audience awareness, a comic routine in which the characters themselves do not seem to catch on to the terms of their predicament, but instead repeat their situation over and over. Even when, at one point further on into the piece, the group riffs on two of the answer words in a sudden burst of creativity that seems to startle them—hole in the ozone layer, hole in the fabric of time, hole in your heart, hole in your head, hole in your pocket, or money in your pocket, secret money, dirty money, easy money, money laundering, blood money—they return to the inevitable fact that money, or hole, are the only words any guesser will ever be able to think of, despite their associative possibilities, despite this tiny breakthrough in thought.

The conditions of thought itself are often foregrounded in the piece: the act of thought. At one point far into the piece, Jerry poses as Rodin's *The Thinker* while wearing the chicken suit. It is as if the conditions of thought are determined to be unattainable or curtailed almost to the point of zero. The "Thinker" is in a silly costume referencing an animal thought to be among the stupidest of animals, one that can run around without any head at all. This near point zero of thought must be conducted in timed conditions, under the pressure of acceleration, with the arbitrarily inserted tick tock sound, the reminder of time, not as passing but as monitoring the time thought is taking, thought that in any case has no possibility of originality.

The brief improvisatory word riffs occur as the contestants devolve into increasing frustration and disappointment. They have become angry at each other for not getting it. By this point it is becoming clear that they need very badly for someone to get the right answer so that they can stop doing this. The excitement affect is gone; they cannot get it back; they cannot restart the game with anticipatory energy—they are wearing out. They are stuck. They begin to skip the steps in between beginning each round, no longer stationing themselves at their mikes, hurrying through the MC's questions in a monotone, interrupting each other, cutting each other off, already running upstage for a swapped garment before the word "swap" has even been spoken. The canned audience laugh and clapping is now completely out of coordinated time with events of the game. Those sounds now seem to happen randomly.

Part of the game all along is that the MC can announce a break, which means that they "do the dance." For the dance they all don a chicken outfit, or don it partially, with or without the head, and do a strange, slowed down version of the "Chicken Dance" with a sort of pained determination, staggered at different spots around the stage, very much in isolation from one another. There are many examples of the "Chicken Dance" on YouTube. It is a popular little dance ditty, party fare, supposed to keep it light, the simplest and most mindless form of happy-making. Here it is increasingly sort-of danced—as the break from "thought," the break from the increasingly aggressive helplessness of the entrapping game.

As the game continues (and gets funnier and funnier for the audience), the desperation of the three grows. There is a slight change in their behavior. They begin to cheat. But neither is this successful in causing someone to win and therefore ending the game. This is because only one of the two, guesser or thinker, is ever willing to cheat. One of them in every case sticks to the rules even when provoked to cheat, encouraged to cheat, helped in every way to cheat. For instance, Claire puts the blindfold on so that it does not cover her eyes, like a child who intends to peer under it. The other two are dying for her to do so, to look at the word on the sign, but it does not seem to occur to her to peek. The same happens when Jerry, who has by this time long since jettisoned the blindfold entirely, will not look at the word even when Richard as thinker has taken his sign ("caravan") over to him and holds it so that Richard can look straight at it. Claire, now MC, and trying to encourage Jerry's cheating, says to Richard, "We want you to be a winner." "Just look and you shall find." "Sometimes the answer is right in front of you." Richard, thinking and thinking, looking straight at the sign, still comes up with "money." Claire prods him; "Are you by chance spelling it c-a-r-a-v-a-n?" Richard responds earnestly, "No, m-o-n-e-y." Claire and Jerry are enraged, Jerry throws the sign in Richard's lap and suggests taking a break. "We'll do the dance."

Now the dance is accompanied by a slow mournful violin. Jerry dances compulsively and alone. They all strip down to underwear and change clothes. Claire is now in a sequined gown, Jerry is in Richard's earlier underwear outfit, and Richard is in Jerry's original wig and suit. They are about to re-boot. The fluorescents fizz and snap with an electric scuzz. They are all in a form of seizure, shaking. Richard, at the mic, is fuming, weird, raging, everyone in agony. They swap. Not even bothering with signs at this point. Now Jerry is at the mic. Claire is the guesser. Jerry, furious, says "Don't open that mouth until you think, Claire." She guesses, wrongly of course, "money." The music vamps. They run around frantically, changing sides again, frantically changing clothes again, faster and faster, escalating, saying all the lines as they run around, abandoning the structure of the game. They lapse into absurdity, or absurd mistakes: "What's your name, Jerry?" "What's Richard's name?" They re-boot. Slow back down. Take another try. Richard goes back to trying to be funny with his sign. He has "sausage" and he turns sideways to the audience with it and inserts it between his legs first as an erection and then sagging. Wink wink. They begin mocking each other for not getting the answer.

In another round, another attempt at cheating, Richard, at the mic, is laughing at Jerry. Jerry is now the thinker. Claire is the guesser. Claire says to Richard as MC that the word is "money." Her absolute certainty that the word Jerry is thinking is "money" becomes increasingly exaggerated. She is trying to hint to Richard to get Jerry to agree that his word is "money." Then this could all be over. Richard says to Jerry, "I'd like to think that the word is 'money.'" "I just need you to confirm that the word is 'money.'" Jerry will not do it. He says "money" is wrong and that the answer is "hole." Richard mocks him: "He thinks it's 'hole'! Where are you getting this stuff Richard? You're not going to believe this but I've just heard back from Claire and you're obviously wrong." Their exhaustion and frustration grow. It is abundantly clear by now that what they want is for this to be over and that they are increasingly aware that they are stuck. They re-boot again.

Now Richard has the sign, "sausage," and he whispers it over and over to Claire as guesser, gradually just saying it right out loud, going up right in her ear and saying it. He mimes cooking sausage. He goes and puts the sausage sign in front of her eyes, points out the word and says "sausage." She giggles at him; she thinks he is being funny. Uncomprehending, and inspired by thinking she has found the answer, she says joyfully, "It's electricity!"

What this performance clearly demonstrates is the absence of a future. The rules lock the performers in place so thoroughly that in this contained space it is not possible to simply call it a day and declare the game finished. They do not seem to be allowed, or to allow themselves, to say, "We don't want to play this anymore. Let's all go home." All they can do is do the "Chicken Dance" and "come back to this later." The only thing that time applies to is the timing of each round. Jerry says, as our experience of the game, this performance, is drawing to a close, presumably leaving them there: "That's all we have time for. I think we should just do the dance. We can come back to this." The violin slows it all down. They do the "Chicken Dance" slowly and grotesquely. Lights lower to just the fluorescent, and go out. Time cannot go on, but only be repeated. They test and test for variations in stasis. It seems that the longer the game is played, the tighter it binds. It becomes more and more painful for the participants. But it will not break. The frenzy grows, but there is no bursting, no outlet. It is a portrait of capitalism akin to Agamben's formulation in which (in Christian theology) what is eternal is infernal.[2]

In a recent talk, Tim Etchells spoke about the constrictions that Forced Entertainment puts on performance in their newer work like *Real Magic*, in which, for instance, the square of green Astroturf cut into the larger area of the stage is an "edge, a container, a boundary in which what happens is obliged to unfold." This talk, performed at the Performance Studies International Conference in Hamburg in June 2017, began with an image of walking with a throng of people in a protest in London. The police placed barricades so that the protestors were contained into one large square and were not allowed to leave. All that energy got stopped, contained. Beside him, sitting on the pavement, a woman began to speak. Her fantasy of a world gone excessively wrong poured from her, a labyrinth of worded sensations, premonitions, fears. It was as if the containment was a political condition in which political energy was not allowed to "move along," in which it was corralled, or incarcerated, into a situation of non-movement from which a future unfolding was removed. Later in the talk Etchells described *Real Magic* as "endless variation rather than change . . . monumental stuckness, glitches, outbursts in contained space . . . correcting, resetting, restarting, and compliance . . . rules are adhered to, energies are held in place, the system holds." The protestors and their political energies were held in the square. Etchells is asking, as he says in a short interview in a video on the making of *Real Magic*, whether change is possible inside a system, inside a structure, or whether the "structure that we are in is so powerful and crushing that we are unable to really change it."[3]

This "structure" that holds *Real Magic*'s game show also translates as the present. As I noted in Chapter 1, scholarly and artistic attention has turned (or returned) to the present as the material with which to attempt to engineer a kind of protesting temporality, a presentism that enables a new negotiation of past and future into a rich mix of the present. This fascination with the present as the only viable site of resistance, as the only possibility in the face of the disappearance of the possibility of modernity's futures, (including revolutionary), is contested by the emergent group of scholars and artists variously identifying as, among other things, accelerationists, xenofemininists, hyperstitionists. For these, in varying ways, the focus on the extended present is a disavowal or lack of recognition of the actual condition of the present, which has become what Armen Avanessian and Suhail Malik call a "time-complex."[4] As editors of a volume called *The Time Complex: Post-Contemporary* and authors of the introductory essay, "The Speculative Time Complex," they suggest that in the present as time-complex, the future is the core mechanism and force. According to the editors, while the left laments the loss of a viable revolutionary future, and imagines the present as the site from which to contest the speculative futurity of contemporary capitalism, they fail to recognize that the future is already here and operational in the present of hypertrophic capitalism. They fail to recognize that the present is "thinning out" under the "forcing of the future."[5]

In Avanessian and Malik's examples we can think, for instance, of algorithms that predict what we will want to buy and the object appears before we even know we want it, offered via our internet searches or Facebook page. Or we can think of preemptive strikes calling an enemy that might have existed in the future into being now. Or we can think of derivatives, in which two parties agree, in the present, to sell a product sometime in the future, for a future price that has been guessed at. For Avanessian and Malik, disturbances to neoliberalism (visions for a future) are only possible through acknowledging the force of the "speculative time complex" in a present that has been thinned out by it, incapacitated as a politically powerful position. So, in contrast to a kind of weak politics that resists the futures of neoliberalism from the position of the so-called present, they call for a "speculative politics that is capable of accelerating the time complex" (futurities as the present) and "introducing a difference to it."[6] This difference is a start of a radically different future in which these theorists are invested.

In one sense we might accuse Forced Entertainment of this leftish dwelling in the present as the only container for political action, one constituted by pushing on variation after variation in the same, so that something might break out of the container, the present. The content of this present, its limitation of language and thought, is not such that the future could be something radically different from it. From another point of view, we might say that *Real Magic* is exemplary of the approach Avanessian and Malik champion. Etchells is staging an extreme present, a present contained in such a tight space, that little spot of Astroturf, that the pressure to explode becomes exponential, searching for something to happen, to become different. In this sense, then, the future is the present, or there is no longer any differentiation possible between the two. Etchells fills this present with the characteristics of the speculative time complex: acceleration, frenzy, betting, cheating, a battle with future loss that is already here. His now is the now of speculative time. We could say he is accelerating it to find a difference in it, like the woman spewing in the stopped and contained demonstration, an acceleration, uncontrolled, of thought and image.

Both of those things may be simultaneously true of *Real Magic*. But what I want to propose is that what happens in *Real Magic* should not be measured entirely according to the more or less vacuous "resistance" practices of presentism, or as something that can be described as an attempt to source from capitalism itself innovations that found strategic imaginings of or logistical plans for a better future. Let me approach this first through the notion of the hold and say that I am interested in *Real Magic* as a kind of hold, an immobility, that produces a creativity like that of kairos. The present of *Real Magic* is perhaps both the present of presentism and of the time complex, but it is also an experiment with energy in the hold. The "structure" that Etchells refers to, the one that may crush us in its interior, might also be the hold, a form of immobility, an extreme pressure of containment. While *Real Magic* is not kairos, like kairos what happens in this hold does not move on, move into a future, hope to move into a future, but holds at a point of creative irresolution. The real magic of the performance is the excess of creative energy circulating, ebbing and flowing, puncturing, punching, exploding, and electrifying in the performances of Richard, Claire, and Jerry. To think of it as energy in the hold is politically evocative, as I think Etchells means it to be when he describes the containerized protest, protest put on hold, put in the hold, as a result of which a different kind of political imagination surfaces.

The "hold" is a term I am loosely borrowing from Stefano Harney and Fred Moten's essay "Fantasy in the Hold." It is a term that for me can be joined to the kairos that I see in my upcoming performance examples. For them, the hold is historically, politically, and economically literal. But it is also a description of a collective state of creative being or existence shared in common with others in the hold. For them, what happens in the hold is "ungraspable,"[7] of "incalculable benefit."[8]

In the negative space of the hold, that is, in the internal hollow, the dark underbelly of the ship, the airless inside of the container or the van, there is movement, an insurgency, a beauty, a music, "flights of fantasy,"[9] a company of visionaries, of touch, of "advent."[10]

Those in the hold are being moved while being kept contained, motionless. Harney and Moten refer to the global movement of those in the hold as "logistics," and they say that modernity begins with logistics, the vast network of moving things, commodities, (especially things that can speak), things moved by others, but unable to move themselves.

Harney and Moten's poetics of the hold has a temporal aspect. Those in the hold were subtracted, and subtracted themselves, from modernity's futural narratives. Moved, held but moving internally, almost imperceptibly, creating insurgent visions, those in the hold—characterized not only as slaves, but all those whose labor was/is not free in the sense that the worker is free to sell his labor power—do not cohere into anything like modernity's proletariat and its vision for the future. Those in the hold are no thing, nothing. The proletariat was "located at . . . a point in the production process from which it had a peculiar view of capitalist totality." But the others "were located at every point, which is to say at no point, in the production process. . . . The standpoint of no standpoint, everywhere and nowhere, of never and to come, of thing and nothing."[11] The proletariat, "commodity labor,"[12] had a position from which it was possible to see that this form of labor must be abolished (to move toward the end of capitalist society and the installation of communism). The proletariat, the wage laborers, are the visible signs of the promise of the overturning of capitalism in the future. They constitute a plan for the future. By contrast, the containerized—labor that is un-free, having no place in the plan, kept in the holds, transported here and there—are a prophecy, an "insurgent prophecy that all of modernity will have at its heart, in its own hold, this movement of things, this interdicted, outlawed social life of nothing," "an absence . . . that it cannot . . . surround."[13] Prophecy here displaces the futural directionality of modernity in the presence of and among those who have chosen "to be the shipped."[14]

Harney and Moten's descriptive writing on what happens in the hold is beautiful, poetic, and elusive, beginning with the "flights of fantasy" that are intrinsic to life in the hold, moments and music that are "emphatically, palpably imperceptible and, therefore, difficult to describe." Those who choose to be the shipped elect to "pay an unbearable cost that is inseparable from an incalculable benefit." In the hold, they write, "we . . . enter again and again the broken world, to trace the visionary company and join it."[15]

The hold, for me, suggests a position from which proceeding into a future is neither possible nor desirable. It is a place of intensely felt creation that does not resolve into a thing that is capturable as an object included in a futural narrative but remains open-ended. What happens, happens there, in the hold. It is visionary in its imagination, its fantasy. It can step over what we ordinarily call the future to a place of alternative being, or alternative or formerly unimaginable insurgencies. It can prophesy and name what does not exist in ordinary temporal, logistical constructions. It can make it felt. There is a "cost" and a "terror"[16] that accompanies its beauty.

Real Magic, I have suggested, is Tim Etchells's experiment with life in the hold, or politics in the hold. It is not the same as Harney and Moten's hold. It is too captured in the webs, the anxieties and fears of the time complex. But it is the sense of the energy or creativity that happens in the hold, the containment, in the immobilization of the hold, that *Real Magic* models and that I have therefore used to set up this chapter on kairos, or the replacement of the future with the to-come. Kairos is a hold, in the sense that it happens on the edge or brink of time. It happens in the hold. It is an energy, an act of creation in the hold, as Cassils's performances and Dreyer's Joan so beautifully show us.

Tiresias

I saw this piece by Cassils in 2013 at Stanford University as part of the Performance Studies International Conference of that year. From the moment I walked into the performance space and saw this person standing, illuminated, nearly naked, up against a sheath of ice—a male Greek torso made of ice—I was acutely aware that there was a time happening here, a time that had nothing to do with the insistently self-conscious wealth of the neoliberal campus that was its environment.[17]

The space was expansive, glossy, black wooden walls that seemed reflective, or as if shimmering. It felt like an encasement outside of time, or isolated from the time outside entirely, like a tomb or a cave. It was in fact something like entering a cave, one whose watery walls were turned glittering by a pillar of illuminated treasure, or crystal, or ice crystal at its center. In entering a cave, time shifts dramatically. There is a feeling of no time or, better, ancestral time—in Quentin Meillassoux's sense of a "reality anterior to the emergence of the human species."[18] As in a cave, time here in this room was marked by the primordial substance, water, dripping, each drop by slow drop amplified electronically. Cassils was illuminated by a powerful shaft of brilliant white light from above, the only source of illumination in the room. The front of Cassils's torso was pressed into the back concave space of an ice sculpture of a classical Greek male torso that was supported by a stand of clear plexiglass. The front of the male ice body, with its internal bubbling, streaming, cracking, turned the front of Cassils's body behind it wavery and fluid. Cassils's heat melted it almost imperceptibly, but actively. The ice seemed as if illuminated from within, glowing.

In this image, Cassils is silent and still. They are in the hold. Their body is arresting and beautiful, muscles curving out and in and expanded everywhere, arms and hands veined like those of David. We are free to move around and look at them from all sides. Cassils's body is built of extreme discipline and attention, through "weight lifting, explosive power training, Muay Thai, traditional boxing, diet manipulation and supplementation," a trans body without hormones or surgery, a trans body that is not "about a crossing from one sex to another, but as a continual becoming, a process-oriented way of being that works in a space of indeterminacy, spasm, and slipperiness."[19] The namesake of the piece, the blind prophet Tiresias, lived for seven years transformed into a woman. As Cassils writes on their website video of the piece, living as both genders was important to his prophesy. But Cassils removes the twofold gender of his experience. Cassils holds at the point of a transformation without resolution. So, here we are, in the dark and glassy cavernous room, this illuminated treasure at the center, with the dripping of melting ice marking time in echoes of eternity, their eyes blinded (with cataracted contact lenses), their skin reddening, watching the force of their determination not to pull away from the ice, performing, as they say, "the resolve required to persist at the point of contact between masculine and feminine."[20] At the Stanford performance the space was so large that we could stay close, make wide circles and arcs around Cassils, stand or sit on the floor. Cassils interacted with no one. Their attention never wavered, all inward concentration, the spectacularly carved face, the hair cut close and severely, the opaque eyes of the future seer who holds at the edge of time, at the void, where being is not moved forward into a resolved state, but names itself as something uncapturable, open ended.

In the first pages of his small volume *Nymphs*, Agamben talks about a work by Bill Viola called *Passions*. It is another image of the hold, an image Agamben calls kairological. The work consists of videos resonant with and borrowing from classical painting, in particular on religious themes. In summer 2017 in Hamburg, at its International Museum of Photography, I saw one of these, taken from *Christ Mocked (The Crowning with Thorns)* by Hieronymus Bosch. In it five contemporary people, four men and one woman, move almost without moving in an almost imperceptible flux of emotions that fold in and out, move between one and the other, the camera held on them from mid-torso up in a cavernous black background, a void, watching hands, intricate details of facial muscles, eyes, the tilt of the heads. At first sight, it is a still image. One has to stay there, with it, to begin to perceive the movement, the flux, the creation, the ebbing and flowing within the image. And then, it is astounding. Agamben writes about the "kairological saturation" of these images, which "imbues them with a sort of tremor that in turn constitutes their particular aura." The images have "charged themselves with time almost to the point of exploding," and show not "images in time but time in images."[21] It is an interior flux of open-ended recombinant gathering and creating. It can also be a description of the interior kairological saturation of Cassils's hold.

It might be more common to describe *Tiresias* as durational or endurance art. There is no question that the piece has much in common with those forms. The pain must be relentless. The discipline not to break away from the ice must be extreme. (Cassils did have the smallest finger signal to an assistant when they had to come down and, I learned after, be taken to another room and wrapped in a heat blanket for a few moments.) Cassils's intention was to stay in the ice until the ice either entirely melted, or fell away, from their body. And yet, the temporality usually associated with durational art is one that demands attention to an extended present held between a past and a future, a then and an after, a present that is to offer, as in the presentism that has been criticized in these pages, the sources of phenomenological, political, and emotional resistance and critique. But kairos gives us something else. It gives us a "musical moment, a moment of advent," to return to Harney and Moten's language for the hold, and, instead of speaking of endurance, we can speak of an acknowledgment of "cost" and "terror" of the hold. It gives us, as Negri writes, an "adventure beyond the edge of time,"[22] the to-come.

For Negri, the future as usually thought is always based on some prior image, or on "a repetition of what has already happened," or on a form of desire, also based on what has originated in the past. The to-come, by contrast, is always a "creative leap, a difference." In kairos, the imagination is at work on being. Kairos is "a truly constitutive operation that situates its creative power on the edge of time."[23] Kairos leans out over the void of time, but does not "precipitate into it."[24] To precipitate into it, to move on, to move on into the future, is to continue what has been in the past through gradual processional revision, if there is any change at all. To hold at the brink is to separate knowledge, knowing, from the temporal succession by which it is normally thought to be verified. Instead, kairos is a temporal condition in which a new being is named. It is "a particular identification of a temporality which expresses new being."[25] It is a temporality in which the naming and the thing named come into being simultaneously, un-adjudicated by temporally stretched processes of knowledge production, an innovation of being.

Just as Negri fashions the to-come as a replacement for the future, he replaces the past with the "eternal." He does not mean eternal in an ahistorical, transcendent sense. He means it in the sense of an "indestructible temporality" in which kairos constitutes everything that has come before.[26] The use of the word "eternal" allows him to situate kairos as always here because the eternal removes the measurement of time according to past and future. He describes conventional versions of the past as "the accumulation of the destruction of physical events," a "continuous genesis of the present," "a dead, finite time distended in duration," the "sedimentation of concluded human events."[27] The eternal, by contrast, is an accumulation of the vibrant creative kairos, "the power of accumulated life," "the vitality which created what happened before us."[28]

I spoke about that room at Stanford, its glossy, watery cave-ness, its ice, the feeling, in the first instance, of the removal of time itself, the summoning of an eternal in which Cassils is here. The idea of the eternal, summoned by the elemental feeling of cave, helps to make this kairos feel eternal in Negri's sense, an indestructible creative temporality, the temporally unmeasurable vitality of being. The ice (although in predictions of the future based on global warming an endangered substance) is also here to stand for the eternal. It is a substance that is always there, somewhere, outside the measure of past and future. The Greek male torso is a fifth-century relic, well within the historical reach of the Greek tragic dramaturgy of the terrible unfolding into the future of a predetermined fate. Turning it into ice turns that past and that temporality into the creativity of kairos, as the ice cracks and breaks and changes in infinite interior variation. Cassils, with the ice torso, creates a twofold hold on the creative brink where ice itself practises kairos. Cassils is not here as an extenuation of past events, not as an intended installment of a better future, and not as an extended present between the two. Cassils turns the past into a substance, ice, which makes visible the internal creativity of the hold.

Although some people who had watched the performance for a while seemed to grow bored or just had other things to do and left, I felt myself to be in the hold, held. Along with some others, I stayed and stayed. Eventually Cassils left the room. It was one of their warm-up breaks that I subsequently learned about. Those of us in the room were alone with the ice. Beautifully illuminated, it bubbled, and cracked, and ran crooked small streams in its own icy interior, the infinite creation of this form interior, or simultaneous, to its immobility, its hold. Suddenly, though, the torso exploded. It just burst apart, shattered, its luminous, refracting fragments blown in all directions, then falling and scattering across the glossy black floor. The room was still illuminated solely by the straight pillar of light focused on the pedestal, but its light had scattered to be present in the glistening dispersal of ice crystals. I found it nearly indescribably beautiful. It was kairological saturation bursting into unforetold being, in an event impossible to predict. When Cassils came through the door a few moments later to encounter this unpredicted explosion, they smiled, pleased, turned around, and closed the door behind them. We were left with that relic of the past, that masculine icon, turned into an exquisite and unnamable being, in the silence, in the cave, in that feeling of the eternal, until we chose to leave.

In another performance of the *Tiresias* project, it was Cassils who was left after the ice. This art work was a video, which I saw as part of an exhibit of their work called *Body of Work* in 2013 at the Ronald Feldman Gallery in New York City. The video was created by collapsing and editing a four-hour performance of *Tiresias* into fifteen minutes. Cassils was careful to specify in emails to me that they consider this a different work of art from the performance. In this installation, the life-size image of Cassils in the ice torso is rear projected onto a piece of floating plexiglass. In the Vimeo sample that Cassils sent to me, Cassils appears only from the waist-up, and many of the images are close-ups of the ice itself so that we watch that beautiful internal innovation in the ice. What stands before us is a trans Tiresias, a figure in whom futurity collapses into an eternal melting, the eternal motion of the structures and fractures of ice crystals continually changing. The ice, as in the performance, is a co-inhabitant of this kairos; the iconic torso, instead of preserving, monumentally, becomes creative. Cassils's Tiresias is the source heat that initiates a gorgeous fractal motion. While it might be possible to argue that the ice melting is ultimately a linear process, the camera's attention is to its hold and to its intricate transformative interior. In the video, the ice does not burst, as in the unpredicted event in the performance, but melts away, leaving Cassils standing alone, an innovation of being.

If kairos gathers the power to name and call the thing named into existence at the same time, Cassils emerges from their twofold hold as a figure of perpetually undecided gender, perpetually at the "point of contact between masculine and feminine," a figure of the to-come, not the future. They are both imaginary and real, a figure of the imagination generated in kairos as it "casts its net over the to come so as to know it, construct it, organize it with power,"[29] and bodily, physically, really that thing itself. That thing which could not have been made by a traversal of gradual revisions from the past into the future. That thing, which, in the language of Harney and Moten, is a no thing, subtracted from narrativity, an "insurgent prophecy," a beauty, a music, a being that, in Cassils's words, again, is made of indeterminacy, spasm, slipperiness. I might call what has happened a "new being" in Negri's language. But I am hesitant about his use of the "new" because (as I have said, I do not think this is true of Badiou's new present) it seems to me still entangled with modernity's narratives, in which the new follows the old, or is a rupture with the old, or is at any rate in relationship to an unfolding in time from past to future. Negri means it as a rupture, independent of this time-scape. However, I think that conceptually, if almost accidentally, it remains embedded in a kind of extrinsic passage of time that he is in general eschewing. I prefer to think of what Cassils has done as closer to Harney and Moten's description of the almost imperceptible motion in the hold. I prefer to think of naming as naming that which only exists in open-endedness—kairos as a way to step over what we ordinarily call the future, to reach out over it, to name what does not exist in ordinary temporal, logistical construction, again, uncapturable.

Cassils's choice of the figure of Tiresias brilliantly draws our attention to the concept of the future, of seeing the future. But in the hold, these prophetic eyes are blind to the future that the accelerationists and others are identifying as our contemporary situation. In the hold, there is the prophesy of the uncapturable or ungraspable thing, not the future betting of derivatives and speculative investment, not contemporary logistics. In the hold, with Cassils, there is a sanctuary from these. I have already discussed briefly the accelerationist descriptions of the contortion of the future into the present that we are undergoing, of speculative capital. But we can elaborate a little further in order to help to emphasize the importance of the sanctuary of the hold and the urgency of its kairological capacity.

In truth, speculative capital is far from new. Speculative capital values what is in the present according to what it may earn in the future, according to what a product, a commodity, might return at a later date. It creates valuations of a commodity based on that commodity's earnings' risks in the future, and figures how to insure it to guarantee its future value. It started with the hold, with logistics, with the slave trade, with the movement of immobilized cargo. Insurance companies were invented for the slave trade as shipping companies and slave merchants betted on sickness, death, or arrival along the circuits of the slave Atlantic. The speculative is a futural logic historically central to capitalism, if gone viral since the 1980s.

The speculative future is imposed in almost all of the places across the globe where some form of capitalism is strategizing and implementing localized operations, and it changes the experience of time itself. With debt structures central to this process, credit and debt become, as Maurizio Lazzarato writes, "capital's way of making/taking time."[30] Time becomes what worries us, the future eats at us, through our uncertain status with regard to the hold that debt has on us, whether or not we will be able to pay it off in time, before more interest is incurred, before, before, before . . . . The future is a threat, installed in our everyday present.

Even the time frame of globalized and unregulated financial markets has changed. These markets are less interested in long-term profit and in investment toward that end than in, as Michel Feher writes, "maximizing the distribution of dividends in the short run."[31] In personal terms, as individuals we attempt to curate our value, not so much, or only, as it will unfold in the future of retirement plans, but now as we try to appreciate and try not to depreciate in social, professional, and even emotional and personal terms. Feher, following Foucault, writes that we become entrepreneurs of ourselves, as "human capital that wishes to appreciate and to value itself and thus allocate its skills accordingly."[32] Neferti X. M. Tadiar writes that "the new financial plan for living requires a daily investment, assessment, and management of one's contributions to 'occupy the kind of time and space once readily conferred on personhood.' "[33] Of course, this situation is unevenly distributed, since, as Tadiar says, not everyone is able to "qualify for the investor model of subjectivity." For Paul Stephens and Robert Hardwick Weston, time is measured or experienced in what they call "chrematistic" terms (having to do with acquiring wealth), that gauge past, present, and future as means of self-appreciation.[34]

And then there is neoliberal prophecy, called previval, that development in the genetics industry which means that those who have the money can bet on a sickness they may get in the future, even though they are currently healthy, and take action in the present, like the removal of a breast, to prevent it. Coleman Nye calls "predictive genetics" the new "divinatory technique." She points out that the combination of the new life-sciences technologies with statistics, the favored tool of divination since the Industrial Revolution, has resulted in an increase in "practices of protection alongside the explosion of registers of potential danger, leading to increased attempts to manage insecure futures."[35]

In the hold with Cassils/Tiresias, divination shares nothing with the speculation on object/bodies, commodity/bodies, betted upon, and self betting selves. It is blind, except, we could say, to the vitality of life in kairos, a volatility, or a conflagration—even a naming of something uncapturable and open-ended, transformation without resolution. I want to stay here, with Cassils and the ice, with this twofold, extraordinary temporal praxis. I do not wish to move on. Divinatory power is at work in the hold and it has me in the hold.

When I entered the room at Stanford, knowing neither the artist nor the name of the work, my first thought was of Joan of Arc, ice instead of fire. The association stayed with me for so long that I turned to the final forty-five minutes of Carl Theodore Dreyer's film, *The Passion of Joan of Arc*. Here is another image of immobilization, another innovation in the hold. Both Joan and Cassils are biological women who have encased themselves in the male, in the one case the classical Greek torso and in the other the armor of the thirteenth-century soldier. Both are trans, in Cassils's definition of persisting at the point of contact between masculine and feminine. Both are encased in elemental and eternal substances, ice and fire; both suffer the cost and terror of their decision to join the visionary company of the hold.

The portion of the film of interest here begins as Joan is brought outside to witness the stake that is being prepared for her burning. She is about to be offered one last time the opportunity to confess her supposed heresy. The imagery of the film is of things hewn from the elements, stone, sky, iron, wood, earth, the pores in skin, everything grainy, white, black. The faces of the peasants, who at this point are enjoying a grotesque carnival, are themselves carved deeply and deeply individually, as if from wood already well worn. Mortality marks all the faces in the film, through the details of wrinkles, blights, fat, and broken teeth. Faces are thrown up full frame against the empty sky. Each frame contains minimal information, the participants in the world's fullness deleted, excised, so that the only thing that seems to exist is this arriving present of Joan's immolation. Throughout what follows there are images of instruments of torture, swinging in the empty air, black and vivid.

Falconetti, in this most famous of performances, is filmed in extreme close-up much of the time. Joan's eyes go to the pyre, to the preparations for her death. An elongated memento mori scene begins as she watches a grave being dug. The camera puts the viewer about a quarter of a way down from the top, inside the grave, as the very end of a shovel appears, tossing dirt. In the distance, seen just barely from below in the pit, an out-of-focus cross rises. As Joan, and we, watch, the next toss of the shovel tosses a human skull to the edge of the grave. The camera moves back and forth from Joan's face as more shovelfuls of dirt land on top of the skull. Her gaze lingers on a patch of untended grass and white wildflowers. Her glance travels back to the skull, which is now, through her eyes, in extreme close-up. We can see its pocked and broken ravages just as white worms begin to emerge from the eye sockets and nose opening. The camera cuts back and forth from Joan's face to the skull and its worms. Joan pulls back, her eyes go up in their sockets. One of the clergy screams "France has never seen such a monster" and threatens her with being burned alive ("the stake awaits you!").

Long seconds go by. Her eyes turn to the church's faces on every side goading her to confession, and then away. Her eyes seem to move between seeing this world, in the time in which these terrible things are happening to her, and seeming to see else-where, in an else-time. She is a young woman. What she is thinking is so clear as she watches worms in the eye sockets and imagines them in hers, as she turns her head just a little to the side and sees the small wildflowers in the grass as what she would lose. The scene illuminates the beginning of something that Joan must do. She is looking at a future, her future. She is thinking of a future, one in which her flesh will be eaten by worms. She has to come to see elsewhere than that future, to see, as she will say later, paradise instead. Instead of the future, the chronology in which her young flesh will be destroyed, she must bring herself to the edge of time. There are tight shots of terrible weapons, black hooked balls, hanging against the terrible white sky. She ends limp and will-less. She is lifted from the chair under her arms and taken to a table where the confession is laid out for her. As if in a trance, she signs her mark, and then her hand (she is illiterate) is guided to write out her name. Her punishment is read: "perpetual imprisonment." It is a consignment to the infernal power and violence of the church, to its infernal chronological time without end. Her eyes widen in shock and horror. It was not what she was expecting.

In the next shots she is inside and her already short hair is being cut to the scalp. As she watches it being swept up, she suddenly cries out, her eyes electric, that she has lied. The clergy gather again. She says that "I denied God to save my life." Again, the long studies of her face, the inner process, the inner decision, the decision for the cost and terror of the hold, the fire.

The sequence of her burning begins as, dressed in a shapeless white felted wool garment, Joan emerges through the low door in the whitewashed walls of the church/castle. The way ahead of her is flanked by soldiers and their upright spears. Joan takes the cross that is given her. She is tied to the stake and, when a guard drops one of the ropes with which he was tying her wrists, she bends to pick it up and hand it back to him. She has chosen. The peasants have gathered. They throng, press up against bars keeping them out, many of them crying as they watch. The church's weapons are everywhere. Immobilized, she prays that her suffering will not be too long, and asks God, "Will I be with you tonight in Paradise?" The fire is lit, and with the smoke bubbling and swirling up toward her, her eyes retreat into what looks like a kind of blindness, a kind of prophesy and making of an uncapturable being, leaning out over the void to paradise. Instead of ice, this time the elemental and eternal substance is fire and the fire like the ice supports kairos. Like the flock of dark birds in that white sky that is one of the last things Joan sees, the swirling smoke rising to the sky carries this being who is no longer the church's heretic, but saint, released from the church's infernal future.

As Joan is burning, the camera turns to pan across the heads and faces of the watching peasants. Weeping, they do not budge. They are held, in the hold, with her. Something about the way that the peasants are immobile alarms the guards. The camera cuts to a guard decked in chains, who seems to have spotted something that turns his face aggressive and mean. Other guards too become uneasy with the peasants' presence. A guard turns over his shoulder, where we now see that many other guards are at the ready, including three with their spears at the top of one of the tall stone walls of the church/castle. The camera cuts to Joan, encased in smoke, enduring, persisting, holding the moment of creative time open, struggling mightily to do so. The camera cuts back to the castle wall. Horrifyingly, a hand reaches out of a small window in the wall and drops, over and over, balls and chains down to the waiting hands of guards below. The church and the army can feel it too—that something is being generated during this time. We see Joan's head, through fire and smoke, at last drop to her chest, and she dies.

At this instant, a peasant man names what has not yet existed: "You have burned a saint!" A guard, hearing this, signals for the soldiers, carrying the hideous weapons that have been amassing throughout the scene, to attack the peasants. In the same instant, the peasants revolt. It is, literally, a fiery riot. They refuse to disperse, to move on. The flame around Joan's corpse seems to grow and grow and spread everywhere. The peasants storm across the bridge, storm through the soldiers. The soldiers are swinging their balls and chains, raining spears down from above, whipping people, hacking at them. As the flames reach their fullness, the guards are beating the last of the peasants back across the moat on a narrow bridge. Sometimes the camera looks up at the blows of the ball and chain as if from the point of view of a person on her back receiving them. Dreyer is clearly concerned to emphasize the unstinting brutality of those in power. The hold was the shelter from the church's violence, which stretches out into the future as it reproduces itself, a chronology that wishes to make itself immune to revolt. Shoving the last of the people off the bridge, the soldiers raise it and run back to the safety of their edifice. As the bridge is raised, the peasants on the other side of the moat do not disperse, but sink to their knees to pray, unwilling, still, to take leave of what has happened. Joan's promise, in this film, is that of the to-come of kairos as a time of revolt against violence and those who have the power who administer to it.

In 2015 Cassils performed a piece at the National Theatre in London called *Inextinguishable Fire*. In this piece they were set on fire. Much of

Cassils's work is directed against the violence enacted on the vulnerable, on the defiant, on those who are uncapturable; we might say on those who are the shipped, especially LGBTQ people. In this piece, they were borrowing from a 1969 film by Harun Farocki (a German filmmaker), which "is all about how impossible it is to effectively depict the horror of napalm on film." They say, "My gesture of self immolation speaks to both the desire for and the impossibility of knowing such horror . . . . I experience the very real human terror of being lit on fire."[36] They say, "You can place yourself in that position [being on fire] and think about what it would be like to be subjected to that violence."[37]

I did not see the performance, but it is beautifully documented on their website, from which I have taken the following quotes. Cassils stood still on stage as the audience entered, naked except for a pair of black underwear. Their body resembles the form it was in for *Tiresias*, the adamant strength, resolve, musculature held in readiness. They are standing on what must be a fireproof pad, black and rectangular, with objects lying neatly behind: the covering and clothing for the fire. From beneath each of their feet a wire stretches at an outward angle from the body downstage and offstage. The performance was based on a Hollywood stunt call "full body burn." A stunt team came on stage and prepared Cassils's body by "laboriously coat[ing] the artist's body in layers of fire resistant long underwear soaked in a freezing solution intended to induce a hypothermic state," a state of being intensely chilled, which prevents the sweat from the body from boiling and burning the skin. A short film on the website documents their shivering and shuddering with cold, what looks like ice being stuffed into the black fabric now encasing them, into the opening left across the face where their eyes and nose are visible. Eventually, they are encased in a white hooded jumpsuit. In the film, there is a whooshing sound like the sound of fire traveling a line to the point of ignition, and a line of text appears which says, "If you inhale, your esophagus will burn," and someone calls out, one of the stage hands, "You're on fire!"

It feels terrifying. The film closes in on their eyes, blue, staring out, again, resolve, seeing elsewhere, not the same as Joan, but like

Joan, dense flames leaping up all around their head. In a still shot it is possible to see the entire body encased in fire. Cassils stands, still, with arms outspread. Fire engulfs them. The camera backs away. It is horrifying, even from the safe distance of the Internet, and even though surely people in the audience also knew this was taking place under highly controlled conditions. It lasts for fourteen seconds when a voice, sounding nearly panicked, barks the order, "Down!" and Cassils throws themselves instantly, spread eagled, to the ground, and stage hands run up, literally run up, with hoses to extinguish the fire.

Immediately after, a film of the stunt, made at another time, is shown on a huge screen on the outside wall of the theatre, where people are stopping to watch. The background is a projection of a spectacular pink/red/black sunset/sky and Cassils, burning, is backed by a tall black rectangular shape. In voiceover, Cassils says that the "fourteen seconds full body burn is extended to fourteen minutes slow motion frame . . . slowing the burn down demands the viewer spends time in a world reduced to fleeting headlines in our Twitter and Facebook feeds." They wish for the watcher to agree to something like the hold.[38]

It was an event in which Cassils brought themselves to the edge of time, in the sense that it could not go any further than the fourteen seconds of the hold. The work invokes the Joan of Dreyer's film, the edge of time, the kairos. Even as carefully controlled as it was, the performance suggests the power of self-immolation as a form of kairos. In July 2017, *The New York Times* featured an article on self-immolation. The gist of the article is that, while the self-immolation of Mohamed Bouazizi, which is credited with starting the Arab Spring, was passionately felt and directed toward political outcome, there is a rash of self-immolations, especially in Bouazizi's Tunisia, that are just desperation and/or acts of escape.[39] Can we look at the Joan scene, at Cassils's choice to stage self-immolation, though, and wonder if these are not forms of fiery initiations of kairological being? Perhaps they are done, in conditions of violence and injustice, as a deliberate way to bring oneself to the edge of time, out of the narrativity of hopeless futures carried out in and through violence, a way to be in the hold, to lean out over the void, bearing the cost and the terror, toward alternative being, indeterminate, slippery, unresolved, unstarvable, untorturable, un-napalmable, a being that will be named in the remaining world as, say, the Arab Spring.

The ways in which we imagine these performances as kairological in turn illuminate the ways we might think about the function of kairos in the world. In a final example, I would like to take a lighter turn in thinking about the creative praxis, kairos, on the edge of time. How can this praxis be materialized in performance as a practice of a kind of communism, or at least communalism, health, and joy? How can we see it, perhaps especially in this way, as an initiation of history?

Beytna

Beytna is a dance piece by the Lebanese choreographer Omar Rajeh and his Beirut-based Maqamat Dance Theatre in collaboration with chorographer/dancers from Japan (Hiroaki Umeda), Belgium (Koen Augustijnen), and Togo (Anani Sanouvi), along with Trio Joubran, a Palestinian group that plays traditional oud music. I saw it in the World Theatre Festival in Hamburg in June of 2017.

I want to relay my experience of it as the practicable, real creativity of the hold as joy. Earlier in this chapter I said that kairos, kairos as a hold, is not necessarily the absence of physical motion. As with the ice, Beytna materializes and makes visible, audible, and olfactory an uncapturable interior movement of the hold, a making of alternative being.

It begins with a table, a very long steel table, like a cooking table in restaurants but much longer. It sits onstage in its own light, lit in lovely, ethereal, gleaming tones. It is set so that it is center stage, perpendicular to the audience. The air is fragrant with mint. On its surface are bowls of vegetables: mint, parsley, cucumber, tomato, onion. The table is mysterious. I wonder if the food is real, or not, if the table is an installation that will be a backdrop for the dance. The table is tantalizing because its other promise is that something will really be made here, with all this good, fresh, healthful food. There is nobody in sight.

When we are finally in our seats, a sixty-something woman in an apron (who we learn later is Rajeh's mother) enters and goes to the center of the table. There, she begins chopping vegetables on a cutting board. For a while, there is just this, her seasoned work, long habits of the mother in the kitchen. Before long, seven men join her at the table. Three of them have carried in ouds, string instruments. Each of them, it becomes apparent, has a station at the table, with cutting board and knives. The bowls of vegetables and herbs, it turns out, have been placed so that each station is supplied with ingredients. The woman, speaking in low tones in Arabic, instructs them on the procedure. They each begin to cut and chop. The smells of mint and parsley grow stronger. The work almost immediately becomes playful among the men, trying to outdo one another's pace, as each grouping of two empties the cut vegetables into a big steel bowl between their stations. They are taking great pleasure in the work and the woman smiles at them and they at her. Everyone is doing their work well, contributing. The oud players and a drummer begin to play. The men begin to move, at first only singly, out into the space. They dance. When they do this, the others swing the table so that it is now upstage and horizontal to the audience. This movement of the table continues throughout the piece, as I will discover, the table itself always dynamic, along with the creation happening upon it.

The dances of each dancer seem discovered from the interior, but always in relation to the oud players. Each dancer seems to be dancing his own feelings, in his own style. When they begin to dance together, sometimes in duets, sometimes all four, what I think I start to recognize is that they are dancing happiness, that the dance is a praxis of concocting, physically naming joy. Much of the work has to do with contact improvisation, ins and outs of connection and transference, between each of the dancers' languages. The food work at the table is continual. Someone in addition to the mother is always there, working. When they are, they watch the dancers, sometimes moving in rhythm with them, unable to contain energy, always smiling, always taking pleasure. Sometimes they shout out. The music is very beautiful, the oud players inter-responsive to the dancers. More than sustenance, this is nourishment, as if, while cooking, they dance sometimes, and in both they are sharing, laughing. It is as if dance is no special thing, no professionalized object, but a part of communal nourishment. They are cooking and dancing a beautiful time into being.[40]

On the wall at the back of the stage, high up, there is a clock, a digital face. Here the clock is ticking off seconds and minutes. It appears to have no meaning, or impact on what is going on. At about twenty-nine minutes it disappears and the stage lights go off. We see a small screen on which is projected a grainy, amateur, black-and-white film that is a documentation of Rajeh's work with the choreographer/dancers in the studio. Rajeh, in the film, stages the work within the work to find a creative praxis of a kairological saturation. When the clock goes on again, after the film, it has all the time been moving forward. We are now at thirty-four minutes. Paradoxically, the forward progression of the clock only marks the elimination of the press of forward moving time. They, and we, are in a hold. There is only dancing, returning to the table and creating the food, dancing, music, exchange, and sharing. I observe that the clock has moved to 40:54; forty minutes have gone by, but I observe this only to note that, most curiously, we are not in the time that is passing.

Obviously, the food preparation is proceeding and is therefore embedded in some kind of passing of time toward some futural event (when the food will be ready and eaten). But as in the other performances I have described in this chapter, futural narrativity is transformed by the medium (earlier, ice and fire) into the sense of making without moving on. The table is suddenly and quickly pushed all the way downstage, horizontal to the audience. This close, and from this angle, with the mother behind it, we can now see clearly the enormous stainless steel bowl of salad that has resulted from each smaller bowl being poured into it. It is magnificent, a beautiful fragrant mixture of greens and reds. We can see also now a huge pot, and baskets of flat bread. The woman begins to pour salt on the salad. Some audience members groan, perhaps appalled by the amount of sodium. But if so they fail to recognize this woman's expertise. Such a salad requires a great abundance of salt, a generosity of salt. She pours great quantities of oil over it, and begins to mix it with her hands. One of the dancers turns the bowl for her. Rajeh steps to a mike and introduces the woman as his mother. He invites us, now, to eat, to share. Even though we might increasingly have expected this outcome, it is still stunning, an act of inordinate generosity. There must be at least 160 people in the theatre. The performance becomes a time of eating and communal sharing, lazy, all the performers ladling, serving, bringing plates to us. We share beans, which turn out to be the content of the huge pot, the bread with Zatar, the salad, and even small cups of Arak. We stand around the stage, talking to strangers. The clock, I notice, has at some point begun to go backward, by seconds only, not minutes. When it gets to zero, it resets to a second less than it reached before. It resets to thirty-seven, then to thirty-six, and so on. Reaching less and less further each time, it resolves to zero. An edge of time. As we finish eating, the dancers begin to make small dances in our midst, a way to clear us back to our seats, I thought, but they do not seem to mind at all when many people stay in the playing area. The lights go out again, the oud players turn to face the small screen which again appears, accompanying the work in the studio that we now see again.

I am reminded of Kentridge's studio, as I discussed in the preceding chapter, the studio as a place of transformation, performance, labor, experiment, over and over. Everything on stage, I said about Kentridge's *Refuse the Hour*, is in interaction with the creativity of the studio and its work on changing time. In Kentridge, one important thing about the studio is that it is not presented as the preparation for the performance event that will unfold from it in the future. It is presented as the constantly generative heart of the performance. While Rajeh's studio work is not explicitly directed to changing time, as is Kentridge's, it performs a similar function, when shown on stage, of insisting on creative praxis not as preparation, not as an anterior factor to a performance that moves into the future as a finished product, but as a willful hold of constant, and in this case joyful and communal, creation.

The studio film ends, and, after a few more dances in the semi-cleared space, the performance is over. The ensemble of dancers, musicians, and mother begin to clean up the table, to clear our plates, even as they invite us to stay and finish up the remaining food, which many people do.

In this chapter, I have only just a few paragraphs ago mentioned my titular concept of the initiation of history. I have, to remind my reader again, defined "to initiate" as the creation of a temporal disorder by means of a temporal innovation in what we understand as history. With such an initiation, history, what is meant by it, becomes the manifestation of a radical human capacity, deployed along with or in the interest of other entities, and found in temporal interstices and innovations, in order to inaugurate universal (belonging to everyone and excluding no one) conditions of justice, recognition, and equal distribution in particular local and/or globally related situations. Kairos, the hold, leaning out over the edge of time, is the creation of such a temporal disorder, an initiation. In each case, in Cassils, in Joan, and in the joyful work of *Beytna*, there is an arrest of the historical processionism that naturalizes the injustices and violences of gendering, that violates practices of communalism, sharing and communism, and turns the possibility for joy into very real anxieties, into the harm of the predation by the future installed in the present through debt, betting and self-valuation, and derivatives. Processional history depends upon the future unfolding in its own image. Here, each example is one of stalling, a hold, withdrawing from proceeding into the future. Each is a kind of production of a radical human capacity to inaugurate universal conditions of justice.

It is not that there is no such thing as what we usually call the future. It is not that nothing will happen subsequently to this time, this historical position, this lifetime that we are in. But what will come can be named through a creative gathering in the hold on the brink of time, where time has not, cannot, move forward; we are at its limit, or our praxis defines itself as taking place at time's limit. We can refuse to go on, although we lean out over what is to come. We can do this without yet knowing what it is, until we tolerate, rejoice in coming to the brink, and holding, or being held there, in an act of creative praxis. This praxis will name a being unforetold by the narratives of a stream of "concluded human events" that rob events of their kairological capacity. It is a matter of how to do the future (as well as the past) in a way that cannot be predicted via the ongoing mechanisms, apparatuses, narratives, violences, exploitations, and oppressions that have named and continue to name and treat peoples, things, and species as subjected.

It is possible to take hope from framing outrage and activism, from insurrection to organizing, against conditions of oppression and immiseration by framing them as new presents. It is possible to live better, to find a better form of existence, of political existence, by learning to think of some situations as penultimate times in which we are changing time. It is possible to feel time gathering rather than moving on. I think of an informal conversation I had in the summer of 2017 with a group of brilliant young Arab scholars who were discussing the necessity of a "new Arabism." Each said that they did not yet know what it was, or would be. Their situation is exemplary of a refusal to predict a future based on a narrative of "concluded human events." Instead, they are willing to hold on the brink, to welcome the brink, the leaning out over the void of time, in order to ensure the creativity of praxis, the naming of unforeseen being. They work in or for the advent. Again, it is not that there will be no "future," but that what comes does not so much come "after" as come, again in Harney and Moten's words, as an advent, a nativity.

It is as if Benjamin's angel holds, that angel facing backward, holding, in the force of the wind blown from the direction of the past, that angel who wishes it could raise the dead it sees, piled in the heap of history's destructions. We can imagine it shaking with the wind whose force, if not withstood, will drive it backward, helplessly, into "progress." Its wing feathers are blowing wildly as it leans into the wind with all the force of the hold, the refusal to be blown away. This could be the praxis of the to-come, the naming of what will be, the angel's back turned against the historicist future with all the strength of its being. It is as if it gathers a saturation of time on the brink in which, perhaps, an uprising of the oppressed will name an altogether different being in the to-come. But in Benjamin, the angel, or this hold, this praxis, is defeated. The outspread wings that give the angel the force and the beauty—the divinity, even—to stand on this brink are also the cause of its defeat. The wind prevents it from folding its wings in to its body and instead holds them in that shape that can only gather the force of the blowing. It must be blown away, backward. Hold out, angel. Do not go. Be like the woman in the image from the riot in Ferguson which I described very early on in this book. The woman holds strong against the winds, facing in the opposite direction from the future, holding, where the insurgency that Benjamin's angel wishes for may be already underway, a new being.

Thoughts in the Time Before the End

There are other forms of the to-come afoot, especially in the genre of speculative philosophy, fiction, and practice. Afrofuturism, or the "black speculative movement," conjures imagined futures, and therefore forms of existence in the present. In her "Introduction to Afrofuturism," Alisha Acquaye quotes Mark Dery, who coined the term in his 1994 book, *Black to the Future*, as saying that this form of existence for black people is independent of "the unreal estate of the future already owned by the technocrats, futurologists, streamliners and set designers—white to the man—who have engineered our collective fantasies."[41] Afrofuturism is a widely varied, definitionally unconcluded and open-ended field for invention and creativity. As in the early Afrofuturism of Sun Ra (not called so at the time), some Afrofuturism imagines futures through blends of African or mythic African forms, an archive, with science fictions and dreams of space and space travel. It is also in some forms a vibrant answer to white control of the sciences and their history of subjugating black bodies. The answers to science include new uses of digital and other technologies. As artist Erin Christovale says, Afrofuturism is taking on an "expansive life."[42] The "Mundane Afrofuturist Manifesto," written by Martine Syms, wittily attacks Afrofuturist icons and ideas, including sci-fi genres or technologically optimistic dreams (like space travel).[43] Her video of Los Angeles-based artists talking about Afrofuturism now features, for instance, Christovale saying that her contemporary version of the genre is "pulling from the past to create a new future but specifically utilizing technology and . . . this hacking mentality of taking something that is not yours and using it and creating something new with that."[44] Speaking of the many black women (and queer black women), who are currently creating Afrofuturist music, moving it into feminist registers and at times using robotic and other alter egos, Alison Pezanoski-Browne quotes Ytasha L. Womack to say that "when an artist creates a new origin, a new story and declares that they are from space, that artist is compelled to create new, boundary-crossing music."[45] Those working in Afrofuturism are, whether creating a future now or speculating on futures to come, decisively separating black existence, and black pasts and presents, from processional history's encompassing force.

Donna Haraway explicitly uses the phrase "to come" throughout her 2016 book, *Staying with the Trouble: Making Kin in the Chthulucene*. Speculative fiction or "speculative fabulation"[46] (which she borrows from Latour) is her preferred genre here, used for the "patterning of possible worlds and possible times . . . gone, here, and yet to come."[47] In this book, which abjures the future telling of the Anthropocene and "Capitalocene,"[48] the narratives of destruction, apocalypse and despair, and of human/technological rehabilitation and restoration projects, she works from what she calls "kainos," seemingly a variant of kairos. She means by it something in which what is usually meant by the future (and the past and present) disappears and is replaced by a sense of beginning and newness. It is also full of resonances of memory, what is inherited, "what might still be."[49] She joins this to the word "chthonic," by which she means the earthly, what is of the earth, both from the past and currently, to create her word that replaces the Anthropocene and the Capitalocene: Chthulucene. The Chthulucene comprises a speculative, polytemporal, long-term work between gumen (humans who work in the soil who are "full of indeterminate genders and genres, full of kinds-in-the making-full of significant otherness")[50] and other beings and species that is predicated on overcoming human-driven historical narrations ("the prick tale of Humans in History").[51]

In her final chapter she presents a speculative fiction of her own called "The Camille Stories: Children of the Compost." It is a tale of five generations of Camilles between 2025 and 2425, during the span of which the population of humans (which Haraway considers an enormous problem) is reduced from ten billion to three billion. This happens through processes in which decisions to have human children are made by collectives who are simultaneously merging with their specified animal counterpart—in Camille's case, the monarch butterfly. Camille herself, through the generations, becomes more and more butterfly, beginning with slight genetic modification after she is born. At first her skin is the color of the monarch caterpillar, then of the chrysalis, and then finally the brilliant colors of the adult monarch. She is also able to taste certain wind-borne chemicals that the monarchs use to find preferred flowers. Camille is born to a "community of compost."[52] Haraway replaces homo with its root "humus" to further emphasize the responsibility of regeneration from and in the complex medium of the species-rich soil, as opposed to air, as the site of male human power. They mix, they make new soil. The communities of compost designate three parents, who practise whatever forms of gender they wish, for each human child, and begin to link the child, and the community, into deep relationships of kin which grow and change, but eventually replace insular human reproductive futurism and temporality. The communities work in the damage, in the trouble, specifically in this case the near-destroyed migratory path of the monarchs, from Canada to Mexico. Kin include indigenous people of Mexico, another community of compost, from whom the butterflies are indispensable for existence, and for whom the butterflies are in actual symbiosis with dead humans. Each Camille teaches the next one all that has been researched, learned, and practised toward regeneration in relations with other entities in the older Camille's lifetime. Upon reaching the age of 15, the second Camille requests implants of monarch antennae on her chin to better join the monarch's tastings. The story goes on through more generations, but we might partially characterize Haraway's aims as reducing ontological purities so that "the critters of the earth were forging planetwide ontological revolutions for making kin."[53] Haraway's feminist speculative philosophy, scholarship, and story-telling are directed then toward a world to come, Haraway's own form of leaning out over the edge of time, where she holds against the force of processional history.

In a book on indigenous temporality, *Beyond Settler Time: Temporal Sovereignty and Indigenous Self-Determination*, Mark Rifkin identifies processional history as an ongoing tool of settler colonialism ("the colonial temporality of unending settler succession and futurity"),[54] in this case of the indigenous people of North America. He writes about the ghost dance, a collection of practices that is sometimes called a religion. It began in 1889 with the vision and a prophecy of a Paiute man, Wovoka, that native peoples could reunite with the dead (of which there were so many by that time) by means of a dance through which that healing moment would be brought closer. It became a dance of solidarity and spiritual strength, a practice of non-violence, among many tribes. It is most commonly known through the massacre at Wounded Knee, when dancers and fleeing people were annihilated, and that some say, in another narrative of finitude, marked the end of indigenous defiance and revolt.

Rifkin, however, speaks of the way in which the ghost dance, and prophesy as a relation to the future, circulates in the work of contemporary native writers "to stage alternative visions of Native pasts and futures than those at play in narratives and enactments of settler time."[55] The appearance of the ghost dance in contemporary texts, Rifkin says, is a "cross-temporal communication" that "exceed[s] the unfolding of a timeline."[56] We might say that it is something like a new present. Rifkin notes that contemporary native writers draw on episodes of revolt and struggle from native pasts and therefore show "how contemporary events replay supposedly superseded dynamics from the past."[57] All of this is a means to escape settler time's freezing of moments of defeat into abject and pitiable finitude as a means to periodize native experience and lay out the path to a native future made in the image of the settler. Instead, ongoing prophetic experience as a relation to the future creates a "to come"[58] of open-ended forms of being for Indigenous (as a complexly identified) peoples.

Each of these scholars' thoughts on temporality, on the future as the to-come, on the past as a collection of vibrant moments of life, of existence in defiance of forces against it, expresses the refusal of finitude that is one of the central concerns of this book. Philosophies of finitude tell tales of defeat, and therefore of the insurmountability of time's march forward. It is encouraging that the phrase to-come, or to come, occurs in each of these works as that leaning out over the edge, which we might call a speculative act. We must mean by speculative a creativity grounded, as it is in each of these cases, in a materiality of work for justice in particular and shared conditions of injustice. We imagine a to-come, but we do not depend upon existing narratives. We hold, and what we name in the hold should be somehow extraordinary, a composition of what is not known yet, but is, and may perhaps now be. Some of us may even prophesy with what Harney and Moten call "insurgent prophesy."[59]

These chapters have attempted to divine in three central temporal concepts work for theatre and performance, or work theatre and performance is already doing. They have been an attempt to heed Agamben's injunction, or prophesy, that the revolutionary task is to change time. I have meant to suggest that we can see in theatre and performance already existing work to change time, and if we wish to can imagine theatre's ongoing work as a medium in which it will be possible to continuously generate changed times. Perhaps paradoxically, given the still mostly spectatorial relationship between audience and performance, this playing with changed times reduces the time we spend watching our time go by, the "time *in which* we are," and allows us the experience of "the time *that* we are,"[60] the time we can take to change time. We are not in the theatre or performance space thereby surrendered to the performance's inevitable ending after we have watched the time go by. The change is an active process, an act of inter-imagination with performance, and a desire to co-produce changed time in order to imagine the initiation of history, perhaps in coordination with insurrection, revolt, activism, and organizing taking place outside the theatre, new presents recognized, and tragic denouements rejected.

Allow me to take up Toussaint Louverture again. In his book on visuality, Nicholas Mirzoeff describes the famous portrait of Toussaint, an engraving from 1802, in which he is astride a rearing horse, one hand on the bridle, the other arm arced behind his head holding a sword. He is dressed in European clothes of the period, a general's clothes. Mirzoeff says that, on the one hand, the portrait shows Toussaint's mastery of bodily techniques Africans were thought incapable of, perhaps a displacement of David's *Napoleon Crossing the Alps*. He is large and foregrounded in the image, while the horse's raised front hooves hover over a miniature fortress and the back hooves are grounded in a plantation. Mirzoeff suggests that this shows that Toussaint is in control of "essential technology of the Atlantic triangle,"[61] dominating both colonialism and slavery. The image contrasts decisively to the commonplace imagery used by abolitionists, in which an African man kneels to the abolitionist begging for his freedom.

Further, the image is syncretic, bringing into one the liberation fighter in his European clothes and the Kreyòl god, Sen Jak, who, in the Voudou religion, is one of the Ogou warriors who fight for justice. Mirzoeff continues by imagining Toussaint as a god, as merging a god and himself, as someone who would "allow [himself] to be possessed by others," by a people in this case. When dancing possessed, the dancer discovers "a new sense of 'self' via this other in a body that the dancer had not previously been at liberty to make available for this or any other purpose."[62] Toussaint, represented in this way, with a god in the self, a warrior of the people and possessed by the people, was a "subaltern hero."[63] However, the revolution began to veer in different directions by 1801, with Toussaint's agricultural policies in particular causing revolt among his ranks, among those who were slaves and who were now laborers being treated almost like slaves. Toussaint, the subaltern hero, becomes instead a leader believing he could sway the people by the force of his will, his vision, his syncretic image. There grew an alternative heroism among the people that was in friction with Toussaint as hero. Mirzoeff says that it was delusional for Toussaint to think that a single figure, a hero, would be good for the people, imposing his will. C. L. R. James criticized Toussaint for precisely this.

But Mirzoeff comments perceptively on the idea of the tragic in relation to the Haitian situation. He begins by criticizing David O. Scott, as I have in these pages, and his theorization of the alleged tragedy of revolution and of Toussaint. Specifically, Mirzoeff disagrees with Scott's argument that revolutionary figures like Toussaint are tragic, and are so because they were "conscripts of modernity"[64] who had no choice but to incorporate themselves into the modernist project. There are some similarities here to both Césaire and C. L. R. James, who each compared Toussaint's situation to Lenin's as someone who had to turn to bourgeoisie techniques to save the new Soviet state. But Mirzoeff suggests that instead of comparing the revolution to the Soviet one, and instead of couching it in European terms by which it becomes tragedy, we return to the people of Haiti and the syncretic iconography of Toussaint. He writes:

> It might be said that Toussaint could no longer ride the *lwa/loi*, the "horse" that had carried him through the revolution. If that horse was the people, it seems that they had thrown him. Present-day accounts of spirit possession emphasize the total exhaustion of the rider when the spirit departs. The people no longer imagined themselves as Toussaint, who could in turn no longer imagine what it was that they needed.[65]

Not tragedy, but exhaustion. Not a European genre, but a people throwing off the one who could no longer imagine what they needed, in order, perhaps, to once again initiate their own history.

Finally, Mirzoeff calls the Paris Commune the last Atlantic revolution, an institution of a "new cultural politics of the 'people.'" Those revolutions broke processional history. He says, "The break can be characterized as a sense of being in history as an actor, rather than . . . a simple follower."[66] He speaks of looking at photographs of the Communards and of seeing in them a conviction that they had history in their hands, that they were, in my terms, initiating history. He says that in the West the act of defying death has passed to theology (resurrection) and out of history. He writes, finally, and again about what he sees in the faces of the Communards, those last practitioners of the Atlantic revolutions of which the Haitian was exemplary: "I am tired of mourning. I would like to know what it would feel like to feel so engaged with history rather than death."[67]

I too am tired of finitude and mourning as a replacement for the vital relationship to temporality that the texts and performances in this book offer. The Cuban sisters who comprise the band Ibeyi chanted as a political refrain, at a performance at Brooklyn Steel in Brooklyn, New York, in November 2017: "We are Immortal! We are deathless!" And luckily there are many on this planet who are changing time to the time *that* they are, initiating history, and so do know what it feels like to feel history instead of death. And luckily, theatre and performance are, actively, particularly operational mediums, ones through which time becomes chronogenetic, through which time is in our hands.

Notes

Chapter 1

1 Agamben, *The Church and the Kingdom*, 40.
2 Ibid., 41.
3 Badiou, *The Rebirth of History*, 41.
4 Shechter's principle training was with Ohad Naharin at Batsheva Dance Company in Tel Aviv (Naharin is the creator of the influential movement language, Gaga), and he reports being also significantly influenced by a short but intense time with the Brussels-based filmmaker and choreographer Wim Vandekeybus, also known for his intensely physical and demanding style.
5 Pfefferman, "Star Choreographer Hofesh Shechter Discusses His 'Political Mother.'"
6 See for instance Danavar "Israeli Hora Dance Performance," 2:35–3:14.
7 Badiou, *Polemics*, 208–9.
8 Kaschl, *Dance and Authenticity in Israel and Palestine*, 50.
9 Ibid., 44.
10 Ibid., 51.
11 Ibid., 57. Sturman herself says, "We adapted the biblical words in a positive way to show exactly what happened on the mountain—not a curse but a triumph [in our time]. Today there is an agricultural settlement atop the mountain, and I find great joy looking up to the green on Gilboa." Ingber, *Seeing Israeli and Jewish Dance*, 120–21.
12 Kaschl, *Dance and Authenticity in Israel and Palestine*, 58.
13 The same year, 1948, is called the *Nakba*, or catastrophe, by the Palestinians, and commemorated as such. For further echoes: In a video from one of the worst situations in the Occupation, the city of Hebron, in a pastiche of folk and pop dancing, we can see the extension of the folk dance on the ground in a situation of actual occupation and violence; the line formation, the foot work, the guns. See Eyal Yablonca, "Israeli Protesters Dancing to Kesha." And we can also see the violent internal contradiction of the folk dance invention/tradition in this brief moment

of performed mimicry by Palestinians in Hebron, which overlay the Israeli Defense Forces (IDF) pop/folk pastiche onto the daily violence of searches and arrests of Palestinians by Israeli soldiers. See ISMPalestine, "Hebron 'Dance Protest.'" These two clips, back to back, are something like the interior life of Shechter's choreography: its torsion, its razors.

14 Kaschl, *Dance and Authenticity in Israel and Palestine*, 63.
15 Agamben, *The Church and the Kingdom*, 26.
16 Ibid., 12.
17 Badiou, *The Rebirth of History*, 90.
18 Ibid., 23.
19 Ibid., 31.
20 Bahr et al., "Cops Face Off with Protestors."
21 Benjamin, "Paralipomena," 392.
22 Benjamin, *The Arcades Project*, 473.
23 Ibid., 392.
24 Linebaugh and Rediker, *The Many-Headed Hydra*, 197.
25 Ibid., 198.
26 Bharucha, *Terror in Performance*, 69.
27 Ibid., 69.
28 This quotation is from my notes taken during the performance.
29 Wickstrom, "The Labor of Architecting," 130.
30 Ibid., 133.
31 Ibid., 134.
32 Lubin, *Geographies of Liberation*, 121.
33 Official Watts Prophets, "Dem Niggers Ain't Playin."
34 Esslin, *Theatre of the Absurd*, 29.
35 TwoPlaysinRep, "A First Look at *Waiting for Godot*."
36 Hemphill, "Black Beans."
37 Agamben, *Infancy and History*, 101–02.
38 Osborne, *The Politics of Time*, 49.
39 Agamben, *Infancy and History*, 99.
40 Benjamin, "Paralipomena," 391.
41 Ibid., 391.
42 Ibid., 396.
43 Ibid.
44 Ibid., 393.

45 Ibid., 394.
46 Ibid.
47 Benjamin, *The Arcades Project*, 406.
48 Benjamin, "Paralipomena," 390.
49 Ibid., 395.
50 Tiedemann, "Dialectics at a Standstill," 942.
51 Benjamin, "Paralipomena," 394.
52 Ibid., 395. The passage I am working from here, XIV in "On the Concept," is ambiguous in that Benjamin uses the revolutionary Robespierre as his opening example of blasting a past out of the continuum of history in that Robespierre retrieved ancient Rome and then turns around and says that, like fashion's retrievals of the past, Robespierre only made that "tiger's leap into the past" in "an arena where the ruling class gives the commands." I think it is that he pushes further from the French revolutionary example to Marx, as an example of the revolutionary who is independent of the ruling class, whereas Robespierre was himself a ruler. The tiger's leap that Marx took as revolution takes place, Benjamin says enigmatically, is "the open air of history." It is as if Marx's revolutionaries simultaneously take the leap into the past, retrieving its energies, and clear an opening in history where the ruling class has been dismantled.
53 Farber, *Philosophy and Melancholy*, 172.
54 Benjamin, "Paralipomena," 397.
55 Benjamin, "Paralipomena," 402.
56 Benjamin, "Paralipomena," 396.
57 Agamben, *The Church and the Kingdom*, 4.
58 Tiedemann, "Dialectics at a Standstill," 944.
59 Ibid.
60 Tiedemann notes that while the dialectical image and dialectics at a standstill are at the core of Benjamin's *Arcades Project*, neither has "achieved any terminological consistency" but remain "iridescent" (942). Benjamin's labor on these terms is highly visible in the *Arcades*. See for instance Convolute N, in which he literally writes and rewrites sentences with only the smallest variations.
61 Quoted in Tiedemann, "Dialectics at a Standstill," 944.
62 Agamben, *The Church and the Kingdom*, 5.
63 Benjamin, *The Arcades Project*, 389.

64 Ibid., 392.
65 Ibid., 463.
66 Osborne, *The Politics of Time*, 140.
67 Badiou, *Logics of Worlds*, 508.
68 Ibid., 510.
69 Ibid., 514.
70 Ross, *The Past Is the Present*, 8.
71 Ibid., 13.
72 Ibid., 6.
73 Ibid., 14.
74 Ibid., 12.
75 Ibid., 7.
76 Osborne, *The Politics of Time*, 145.
77 Ross, *The Past Is the Present*, 15.
78 Ibid., 13.
79 Ibid., 13.
80 Ibid., 14.
81 Isherwood, "From Chile, 'Escuela,' a Drama About Education."
82 Badiou, *The Rebirth of History*, 15.
83 Wagner, *Shakespeare, Theatre, and Time*, 4.
84 Ibid., 12.
85 Osborne, *The Politics of Time*, 47.
86 Wagner, *Shakespeare, Theatre, and Time*, 32.
87 Ibid., 16.
88 Ibid., 14.
89 Ibid., 17.
90 Latour, "Trains of Thought," 173.
91 Ibid., 174.
92 Ibid., 176.
93 Wagner, *Shakespeare, Theatre, and Time*, 30.
94 Ibid., 31.
95 Blau, *Take Up the Bodies*, 83.
96 Luckhurst and Morin, "Introduction," 1.
97 Badiou, *Wittgenstein*, 93.
98 Heathfield, *Out of Now*, 13.

99 Ibid., 50.
100 Ibid., 22.
101 Ibid.
102 Ibid., 21.
103 Ibid., 20.
104 Etchells, "Out of Now."
105 Heathfield, *Out of Now*, 23.
106 Jessi, *Spartakus*, 46.
107 Ibid., 53.
108 Wilder, *Freedom Time*, 2.
109 Ibid., 39.
110 Ibid., 43.
111 Ibid., 45.
112 Darwish, *Unfortunately, It Was Paradise*, 140.
113 Ibid., 17.
114 Ross, *The Past Is the Present*, 7.
115 Ibid., 12.
116 Palladini, "The Weimar Republic and Its Return," 19.
117 Ibid., 20.
118 Ibid., 30.
119 Nyong'o, "The Scene of Occupation," 141.
120 Phelan, "On the Difference Between Time and History," 117.
121 Loveless, "The Materiality of Duration: Between Ice and Water Time," 134.
122 Schneider, *Performing Remains*, 89.
123 Ibid., 182.
124 Rockhill, "The Revolution Never Sleeps."
125 Lowe, *Intimacies of Four Continents*, 175.
126 Césaire, *The Tragedy of King Christophe*, 40.
127 See William Shakespeare, *King Richard the Second*, Act V, Scene 5.
128 Butler, *Precarious Life*, 20.
129 Ibid., 31.
130 Agamben, *The Time that Remains*, 68 (italics in original).
131 Kentridge, *The Refusal of Time*, 157.
132 Agamben, *Nymphs*, 4.

Chapter 2

1 Mirzoeff, *The Right to Look*, 70.
2 Stephens, *Black Empire*, 210.
3 Nesbitt, "From Louverture to Lenin," 130.
4 Mirzoeff, *The Right to Look*, 78.
5 Ibid.
6 Ibid., 79.
7 Scott, *Omens of Adversity*, 6.
8 Ibid., 2.
9 Glick, *The Black Radical Tragic*, 5.
10 Wilder, *Freedom Time*, 9.
11 Walsh, *Free and French in the Caribbean*, 111.
12 Ibid., 125.
13 Ibid., 133.
14 Ross, *Communal Luxury*, 91.
15 Castelluci et al., *Theatre of Socìetas Raffaello Sanzio*, 30.
16 James, *Toussaint Louverture*, 10 (quoted in "Introduction").
17 Stephens, *Black Empire*, 214.
18 West and Martin, "Contours of the Black International: From Toussaint to Tupac," 15.
19 Dawson, "The Rise of the Black Internationale," 6.
20 Stephens, *Black Empire*, 130.
21 James, "Introduction," 9.
22 Grimshaw, "C. L. R. James: A Revolutionary Vision for the Twentieth Century," 7.
23 Lubin, *Geographies of Liberation*, 72.
24 Dawson, "The Rise of the Black Internationale," 9.
25 Grimshaw, "C. L. R. James: A Revolutionary Vision for the Twentieth Century," 5.
26 Ibid.
27 Kelley, *Freedom Dreams*, 16.
28 Grimshaw, "C. L. R. James: A Revolutionary Vision for the Twentieth Century," 6.
29 Nesbitt, "From Louverture to Lenin," 133.

30 Grimshaw, "C. L. R. James: A Revolutionary Vision for the Twentieth Century," 7.
31 Quoted in James, *The Black Jacobins*, 86.
32 Wilder, *Freedom Time*, 22.
33 James, *The Black Jacobins*, 47.
34 Ibid., 57.
35 Ibid., 56.
36 Mirzoeff, *The Right to Look*, 67.
37 Ibid., 68.
38 Ibid., 69.
39 Ibid., 70.
40 James, *Toussaint Louverture*, 53.
41 Ibid., 54.
42 Ibid.
43 James, *The Black Jacobins,* 86.
44 Badiou, *The Rebirth of History*, 66.
45 Ibid., 70. (Italics are his.)
46 James, *Toussaint Louverture*, 58.
47 Ibid., 54.
48 James, *The Black Jacobins*, 20.
49 Ibid.
50 James, *Toussaint Louverture*, 55.
51 See for example, https://thelouvertureproject.org/index.php?title=Boukman (accessed August 6, 2017).
52 James, *Toussaint Louverture*, 56.
53 James, *The Black Jacobins*, 25.
54 James, *Toussaint Louverture*, 63.
55 Frey, "The American Revolution," 64.
56 Nesbitt, "From Louverture to Lenin," 130.
57 Wilder, *Freedom Time*, 9–10.
58 Buck-Morss, *Hegel, Haiti, and Universal History*, 32 (footnote).
59 James, *The Black Jacobins*, 171.
60 Badiou, *Logics of Worlds*, 51.
61 Ibid., 52.
62 Nesbitt, *Caribbean Critique*, 11.
63 Ibid.

64 Badiou, *Logics of Worlds*, 64.
65 Badiou, *The Rebirth of History*, 68–69.
66 Badiou, *Logics of Worlds*, 64.
67 Ibid., 67.
68 Ibid., 53.
69 Ibid., 55.
70 Ibid., 56.
71 Ibid., 59.
72 Ibid., 60.
73 Ibid., 63.
74 James, *Toussaint Louverture*, 77.
75 Ibid., 78.
76 Ibid., 79.
77 Nesbitt, "From Louverture to Lenin," 130.
78 James, *Toussaint Louverture*, 58.
79 In the 1967 version of this play, *The Black Jacobins*, James concentrates precisely on the colliding revolutionary decisions that put Dessalines and Toussaint into a final and death-dealing (to Toussaint) battle. Dessalines becomes, as opposed to the 1934 play in which he acts in Toussaint's name, the one who actually betrays him to the French in order to remove him from the scene so that he can enact his own revolutionary vision: a purge of the remaining whites, a declaration of independence, and the crowning of himself as Emperor.
80 James, *Toussaint Louverture*, 65.
81 Ibid., 67.
82 Ibid.
83 James, *The Black Jacobins*, 124.
84 Ibid., 125.
85 James, *Toussaint Louverture*, 68.
86 Badiou, *The Rebirth of History*, 70.
87 James, *Toussaint Louverture*, 69.
88 Ibid., 70.
89 Nesbitt, "From Louverture to Lenin," 130.
90 Glissant, *Monsieur Toussaint: A Play*, 87.
91 Badiou, *Logics of Worlds*, 59.
92 James, *Toussaint Louverture*, 127.
93 Ibid., 122.

94 Ibid., 118.
95 Ibid., 117.
96 Glick, *The Black Radical Tragic*, 140.
97 Ibid., 158.
98 James, *The Black Jacobins*, 291.
99 Ibid., 288.
100 Ibid., 291.
101 Ibid., 292.
102 Lowe, *Intimacies of Four Continents*, 171.
103 Wilder, *Freedom Time*, 191–92.
104 Ibid., 192.
105 Ibid., 193.
106 Ibid., 194.
107 Nesbitt, "From Louverture to Lenin,"134.
108 Wilder, *Freedom Time*, 21–22.
109 Ibid., 33.
110 Ibid., 23.
111 Nesbitt, "From Louverture to Lenin,"134.
112 Césaire, *The Tragedy of King Christophe*, xii–xiii (from "Introduction").
113 Ibid., l (from "Introduction").
114 Wilder, *Freedom Time*, 16.
115 Ibid., 14.
116 Césaire, *The Tragedy of King Christophe*, 9.
117 Ibid., 14.
118 Ibid., 16.
119 Ibid., 104.
120 Ibid., 16–17.
121 Ibid., 18.
122 Ibid., 20.
123 Ibid., 22.
124 Ibid., 23.
125 Ibid., 36.
126 Ibid., 86.
127 Ross, *Communal Luxury*, 16–17.
128 Ibid., 50.
129 Césaire, *The Tragedy of King Christophe*, 60.
130 Ibid., 61.

131 Ibid., 62.
132 Walsh, *Free and French in the Caribbean*, 146.
133 Ibid., 144.
134 Ibid., 137.
135 Ibid., 130.
136 Césaire, *The Tragedy of King Christophe*, 46.
137 Ibid., 49.
138 Wilder, *Freedom Time*, 200.
139 Césaire, *The Tragedy of King Christophe*, 51.
140 Ibid., 55.
141 Ibid., 65.
142 Ibid., 33.
143 Ibid., 34.
144 Ibid., 36.
145 Ibid., 37.
146 Ibid., 38.
147 Ibid., 39.
148 Ibid., 40.
149 Badiou, *Logics of Worlds*, 55.
150 Ibid., 63.
151 Césaire, *The Tragedy of King Christophe*, 25.
152 Ibid., 26.
153 Ferrer, *Freedom's Mirror*, 277.
154 Césaire, *The Tragedy of King Christophe*, 40.
155 Ferrer, "Dark Specters and Black Kingdoms."
156 Ferrer, *Freedom's Mirror*, 277.
157 Ferrer, "Dark Specters and Black Kingdoms."
158 Césaire, *The Tragedy of King Christophe*, 91.
159 Ibid., 92.
160 Ibid., 97.

Chapter 3

1 Badiou, *Logics of Worlds*, 270 (Badiou is quoting Spinoza).
2 Ibid., 268.

3 Wagner, *Shakespeare, Theatre, and Time*, 30.
4 Kentridge, "MetCollects," 3:23.
5 Agamben, *The Time that Remains*, 68.
6 Ibid., 68.
7 Ibid., 65.
8 Ibid., 62. Agamben consistently works in theological language and concepts. While many people find it hard to separate the theology from the thought and find the (political) thought overdetermined by the Judeo/Christian/European foundations of the theology, and to some extent I think they are correct, I nevertheless argue that the theological thinking (sans any "belief") permits a foundational rethinking of time in Agamben's hands that is invaluable.
9 Ibid., 61.
10 Ibid., 62.
11 Ibid., 63.
12 Ibid., 62.
13 Ibid., 64.
14 Ibid.
15 Ibid., 66.
16 Ibid., 67.
17 Schneider, *You Are Nowhere*, 111.
18 Ibid., 89.
19 Ibid., 91.
20 Ibid., 92.
21 Ibid., 95.
22 Agamben, *The Church and the Kingdom*, 26.
23 During, "Elie During. Floating Time. 2013," 32:00.
24 Schneider, *You Are Nowhere*, 100. In the published text, this quotation from Wallace is only provided in fragments. Much of it is blocked out. I am quoting from Wallace from googling the portions of the text in Schneider in order to fill out what Schneider actually said in performance.
25 Ibid., 101.
26 Ibid., 102.
27 Kershaw, *Theatre Ecology*, 218.
28 Schneider, *You Are Nowhere*, 99.

29 Ibid., 102.
30 Ibid., 104.
31 Ibid., 110.
32 Ibid., 111.
33 During, “Elie During. Floating Time. 2013,” 31:00.
34 Agamben, *Infancy and History*, 43.
35 Ibid., 46.
36 Adams, “XXV: The Dynamo and the Virgin (1900),” paragraphs 1, 3, 4, 6, and 8.
37 Schneider, *You Are Nowhere*, 92.
38 Perkinson, *The Alcoholism and Drug Abuse Patient Workbook*, 39.
39 Kentridge, *The Refusal of Time*, VII (Refuse the Hour Lecture inset).
40 Kentridge, “MetCollects,” 3:21.
41 Kentridge, *The Refusal of Time*, 157.
42 Ibid., IV.
43 Ibid., XII.
44 Ibid., XIII.
45 Ibid., 313.
46 Kentridge, *Six Drawing Lessons*, 105.
47 Kentridge, *The Refusal of Time*, VII.
48 Kentridge, *Six Drawing Lessons*, 124–25.
49 Ibid., 49.
50 Ibid., 51.
51 Latour, “Trains of Thought,” 185.
52 Kentridge, *Six Drawing Lessons*, 51–52.
53 Kentridge, *The Refusal of Time*, XIV.
54 Ibid., V.
55 Ibid., X.
56 Ibid., XI.
57 Kentridge, *Six Drawing Lessons*, 108.
58 Ibid., 117.
59 Ibid., 161.
60 Ibid.
61 MyEarbot, “Simultaneity,” 1:44.
62 Here we are of course before the Holocaust trains and the Deutsche Reichsbahn railways system: the full expansion of the railroad’s genocidal function.

63 Schneider, *You Are Nowhere*, 98.
64 Latour, "Trains of Thought," 175.
65 Ibid.
66 Ibid., 176.
67 Ibid., 175.
68 Ibid., 176.
69 Ibid.
70 There is a fascinating other version of Latour's tale in Nathanial Hawthorne's story "The Celestial Railroad." Here Hawthorne describes exactly the smooth trip of the intermediary and the entity-rich experience of the mediator. Hawthorne is writing a kind of Enlightenment update to *The Pilgrim's Progress*, in which there is now a train that glides effortlessly along the pilgrimage path toward the Celestial City, transporting its passengers without event, luggage safely stowed underneath, through all the thickets that formerly made it so hard to reach the City. The passenger who narrates the story describes seeing, at several points, two pilgrims still walking on foot, with their heavy burdens on their back. The pilgrims are mocked. In the end, though, it is only the two walking pilgrims who make it into the city. Hawthorne's depiction of Vanity Fair, where his passenger does disembark, could be aptly applied to twenty-first century capitalist meccas.
71 Latour, "Trains of Thought," 177.
72 Ibid., 178.
73 Ibid., 181.
74 Ibid., 178.
75 Kentridge, *The Refusal of Time*, 31.
76 Kentridge, *The Refusal of Time*, XIII and XIV (of inset).
77 Latour, "Trains of Thought," 182.
78 Kentridge, *Six Drawing Lessons*, 67.
79 Latour, "Trains of Thought," 178.
80 Kentridge, *The Refusal of Time*, XV (of inset).
81 Latour, "Trains of Thought," 179.
82 Ibid., 186.
83 Kentridge, *The Refusal of Time*, XVI (of inset).
84 Ibid.
85 Berlant, *Cruel Optimism*, 199.

86 Ibid., 169.
87 Ridout, *Passionate Amateurs*, 5.
88 Ibid., 63.
89 Ibid., 60.
90 Buck-Morss, *The Dialectics of Seeing*, 265.
91 Ibid., 263.
92 Ibid., 265.
93 Benjamin, "Program for a Proletarian Children's Theater," 202.
94 Ibid., 203.
95 Ibid., 203–4.
96 Ibid., 204.
97 Ibid., 206. (Italics in original.)
98 Gob Squad, "Before Your Very Eyes," 0:45.
99 Castellucci, "On the Concept of the Face," 3:04.
100 Agamben, *Infancy and History*, 111.
101 Ibid., 101.

Chapter 4

1 Bissell and Fuller, *Stillness*, 8.
2 Agamben, *The Church and the Kingdom*, 164.
3 Etchells, "The Making of Real Magic."
4 Avanessian and Malik, "The Speculative Time Complex," 7.
5 Ibid., 34.
6 Ibid., 48.
7 Harney and Moten, "Fantasy in the Hold," 94.
8 Ibid., 95.
9 Ibid., 94.
10 Ibid.
11 Ibid., 93.
12 Ibid., 92.
13 Ibid., 93.
14 Ibid., 94.
15 Ibid.
16 Ibid., 95.

17 Portions of this writing on Cassils have previously appeared in *TDR: The Drama Review* 58, no. 4 (Winter 2014), 46–55.
18 Meillassoux. *After Finitude*, 10.
19 Cassils, "Bashing Binaries."
20 Ibid.
21 Agamben, *Nymphs*, 4.
22 Negri, *Time for Revolution*, 152.
23 Ibid., 161.
24 Ibid., 153–54.
25 Ibid., 154.
26 Ibid., 165.
27 Ibid., 164.
28 Ibid., 165.
29 Ibid.
30 Lazzarato, "Grasping the Political in the Event," 156 (PDF).
31 Feher, "Self-Appreciation," 30.
32 Ibid., 30–31.
33 Tadiar, "Life-Times of Disposability," 21.
34 Stephens and Weston, "Free Time," 141.
35 Nye, "Cancer Previval," 109.
36 Cassils, *Inextinguishable Fire*, 1:12–1:56.
37 Heyman, "Cassils: Transgender Artist Goes to Extremes."
38 Cassils, *Inextinguishable Fire*, 1:56–3:00.
39 Blaise, "Self-Immolation."
40 There is no doubt a decisive feminist critique to be made here (all male dancers, the one female a cook and a mother), but I will ask forbearance here as I leave that to others and keep my focus on the reading of temporality.
41 Acquaye, "Black to the Future."
42 Christovale, "The Mundane Feminist Manifesto," 7:35.
43 Syms, "THE MUNDANE AFROFUTURIST MANIFESTO."
44 Christovale, "The Mundane Feminist Manifesto," 8:19.
45 Pezanoski-Browne, "Black to the Future."
46 Haraway, *Staying with the Trouble*, 42.
47 Ibid., 31.
48 Ibid., 5.

49 Ibid., 2.
50 Ibid., 11–12.
51 Ibid., 40.
52 Ibid., 138.
53 Ibid., 162.
54 Rifkin, *Beyond Settler Time*, 142.
55 Ibid., 130.
56 Ibid., 131.
57 Ibid., 133.
58 Ibid., 141.
59 Harney and Moten, "Fantasy in the Hold," 93.
60 Agamben, *The Time that Remains*, 68. Please note also that I am not suggesting that dissolving the boundary between audience and performer, in and of itself, does anything to change time.
61 Mirzoeff, *The Right to Look*, 109.
62 Ibid.
63 Ibid., 110.
64 Ibid., 113.
65 Ibid.
66 Ibid., 186.
67 Ibid., 187.

Bibliography

Acquaye, Alisha. "Black to the Future: OkayAfrica's Introduction to Afro-Futurism." *Okayafrica*, International Edition, July 10, 2017. http://www.okayafrica.com/culture-2/african-future-okayafrica-introduction-afrofuturism/ (accessed August 13, 2017).

Adams, Henry. "XXV the Dynamo and the Virgin 1900." In *The Education of Henry Adams*. http://www.bartleby.com/159/25.html (accessed August 7, 2017).

Agamben, Giorgio. *The Church and the Kingdom*. Translated by Leland De La Durantaye. Photographs by Alice Attie. London, New York, and Calcutta: Seagull Books, 2012.

Agamben, Giorgio. *Infancy and History: On the Destruction of Experience*. Translated by Liz Heron. London and New York: Verso Books, (1978) 2007.

Agamben, Giorgio. *Nymphs*. Translated by Amanda Minervini. London, New York, and Calcutta: Seagull Books, 2013.

Agamben, Giorgio. *The Time that Remains: A Commentary on the Letter to the Romans*. Translated by Patricia Dailey. Stanford, CA: Stanford University Press, 2005.

Avanessian, Armen, and Suhail Malik. "The Speculative Time Complex." In *The Time Complex: Post-Contemporary*. Edited by Armen Avanessian and Suhail Malik, 7–56. Miami, FL: NAME Publications, 2016.

Badiou, Alain. *Logics of Worlds: Being and Event II*. London and New York: Continuum, 2009.

Badiou, Alain. *Polemics*. London: Verso Press, 2006.

Badiou, Alain. *The Rebirth of History: Times of Riots and Uprisings*. London and New York: Verso Press, 2012.

Badiou, Alain. *Wittgenstein's Antiphilosophy*. London and New York: Verso Press, 2011.

Bahr, Kenny, Pearl Gabel, Corky Siemaszko, and Rich Shapiro. "Cops face off with protesters in Ferguson, MO., as midnight curfew is imposed; 7 arrested, 1 shot." *Daily News*, August 17, 2014. http://www.nydailynews.com/news/national/chaos-erupts-ferguson-day-relative-calm-article-1.1905633 (accessed August 2, 2017).

Benjamin, Walter. *The Arcades Project.* Translated by Howard Eiland and Kevin McLaughlin. Cambridge, MA and London: The Belknap Press of Harvard University Press, 1999.

Benjamin, Walter. "Paralipomena to 'On the Concept of History.'" In *Selected Writings*, Volume 4, 1938-1940. Edited by Michael W. Jennings. Cambridge, MA: Belknap Press, 2006.

Benjamin, Walter. "Program for a Proletarian Children's Theater." In *Selected Writings*, Volume 2, Part 1, 1927-1930. Cambridge, MA and London: The Belknap Press of Harvard University Press, 1999.

Benjamin, Walter. "Theses on the Philosophy of History." In *Selected Writings*, Volume 4, 1938-1940. Edited by Michael W. Jennings. Cambridge, MA: Belknap Press, 2006.

Berlant, Lauren. *Cruel Optimism*. Durham and London: Duke University Press, 2011.

Bharucha, Rustom. *Terror in Performance*. London and New York: Routledge, 2014.

Bissell, David, and Gillian Fuller, eds. *Stillness in a Mobile World*. London and New York: Routledge, 2011.

Blaise, Lilia. "Self-Immolation, Catalyst of the Arab Spring, Is Now a Grim Trend." *The New York Times*, July 9, 2017. https://www.nytimes.com/2017/07/09/world/africa/self-immolation-catalyst-of-the-arab-spring-is-now-a-grim-trend.html (accessed August 13, 2017).

Blau, Herbert. *Take Up the Bodies*. Champaign: University of Illinois Press, 1982.

Buck-Morss, Susan. *The Dialectics of Seeing: Walter Benjamin and the Arcades Project*. Cambridge, MA: The MIT Press, 1991.

Buck-Morss, Susan. *Hegel, Haiti, and Universal History*. Pittsburgh: University of Pittsburgh Press, 2009.

Butler, Judith. *Precarious Life: The Powers of Mourning and Violence*. London and New York: Verso Press, 2004.

Cassils. *Inextinguishable Fire*. Spill Festival of Performance Vimeo. http://cassils.net/portfolio/inextinguishablefire/ (accessed August 13, 2017). 0-3: 42.

Cassils, Heather. "Bashing Binaries – Along with 2000 Pounds of Clay." *The Blog*, Huffpost, February 2, 2016. http://www.huffingtonpost.com/heather/bashing-binaries-along-with-2000-pounds-of-clay_b_3861322.html?view=print&comm_ref=false (accessed August 13, 2017).

Castelluci, Claudia, Romeo Castellucci, Shiara Guidi, Joe Kelleher, and Nicholas Ridout. *The Theatre of Sociètas Raffaello Sanzio*. London and New York: Routledge, 2007.

Castelluci, Romeo. "On the Concept of the Face, Regarding the Son of God, *Sociètas Raffaello Sanzio*." Pacitti Company. (Vimeo). https://vimeo.com/22739969 (accessed August 7, 2017).

Césaire, Aimé. *The Tragedy of King Christophe*. Translated and with an Introduction by Paul Breslin and Rachel Ney. Evanston, IL: Northwestern University Press, 2015.

Christovale, Erin. "The Mundane Afrofuturist Manifesto." https://www.youtube.com/watch?v=otUJvQhCjJ0 (accessed July 2, 2017).

Danavar. "Israeli Hora Dance Performance." https://www.youtube.com/watch?list=RD02UmY2s5U4D1M&v=AyRg1UIUkjI&NR=1&feature=endscreen (accessed August 2, 2017).

Darwish, Mahmoud. *Unfortunately, it was Paradise*. Translated and edited by Munir Akash and Carolyn Forché with Sinan Antoon and Amira El-Zein. Berkeley and Los Angeles: University of California Press, 2003.

Dawson, Ashley. "The Rise of the Black Internationale: Anti-Imperialist Activism and Aesthetics in Britain during the 1930s." *Atlantic Studies* 6, no. 2 (2009): 159–74.

During, Simon. "Floating Time." The European Graduate School Video Lecture. August 30, 2013. https://www.youtube.com/watch?v=1UwblvcMUkI (accessed August 3, 2017).

Esslin, Martin. *The Theatre of the Absurd*. New York, NY: Vintage Books, 2004.

Etchells, Tim. "The Making of Real Magic." https://www.google.com/search?q=Tim+Etchells+Real+Magic&source=lnms&tbm=vid&sa=X&ved=0ahUKEwi9kMqrhq_XAhVSziYKHQzdAewQ_AUICigB&biw=1103&bih=612&dpr=2 (accessed November 8, 2017).

Etchells, Tim. "Out of Now." http://timetchells.com/out-of-now/ (accessed November 8, 2017).

Farber, Ilit. *Philosophy and Melancholy: Benjamin's Early Reflections on Theater and Language*. Stanford, CA: Stanford University Press, 2013.

Feher, Michel, "Self-Appreciation; or, the Aspirations of Human Capital." *Public Culture* 21, no. 1 (2009): 21–41.

Ferrer, Ada. "Dark Specters and Black Kingdoms: An Interview with Historian Ada Ferrer." *Public Archive*, February 6, 2015. http://thepublicarchive.com/?p=4497 (accessed July 19, 2017).

Ferrer, Ada. *Freedom's Mirror: Cuba and Haiti in the Age of Revolution*. New York, NY: Cambridge University Press, 2014.

Frey, Sylvia. "The American Revolution and the Creation of a Global African World." In *From Toussaint to Tupac: The Black International since the Age of Revolution*. Edited by Michael O. West, William G. Martin, and Fanon Che Wilkins, 47–71. Chapel Hill, NC: University of North Carolina Press, 2009.

Glick, Jeremy Matthew. *The Black Radical Tragic: Performance, Aesthetics, and the Unfinished Haitian Revolution*. New York and London: New York University Press, 2016.

Glissant, Edward. *Monsieur Toussaint: A Play*. Three Continents Press, 2005.

Grimshaw, Anna. "C.L.R. James: A Revolutionary Vision for the Twentieth Century." In *The C.L.R. James Reader*. Edited by Anna Grimshaw. Oxford, UK and Cambridge, MA: Blackwell, 1992.

Haraway, Donna. *Staying with the Trouble: Making Kin in the Chthulucene*. Durham and London: Duke University Press, 2016.

Harney, Stefano, and Fred Moten. *The Undercommons: Fugitive Planning and Black Study*. Brooklyn: Autonomedia, 2013.

Heathfield, Adrian. *Out of Now: The Lifeworks of Tehching Hsieh*. London and Cambridge, MA: The MIT Press, 2009.

Hemphill, Essex. "Black Beans." June 29, 2013. https://zocalopoets.com/2013/06/29/essex-hemphill-we-keep-treasure-any-king-would-count-as-dear-poems-of-lust-poems-of-tenderness/ (accessed August 2, 2017).

Heyman, Stephen. "Cassils: Transgender Artist Goes to Extremes." *The New York Times*, International Arts, November 18, 2015. https://www.nytimes.com/2015/11/19/arts/international/cassils-a-transgender-artist-goes-to-extremes.html?_r=0 (accessed August 13, 2017).

Ingber, Judith Brin, ed. *Seeing Israeli and Jewish Dance*. Detroit: Wayne State University Press, 2011.

Isherwood, Charles. "From Chile, 'Escuela', a Drama About Education." *The New York Times*, January 15, 2016. http://www.nytimes.com/2016/01/16/theater/review-from-chile-escuela-a-drama-about-education.html?_r=0 (accessed January 6, 2017).

ISMPalestine. "Hebron 'Dance Protest' Against Israeli Soldiers and Settlers (10th July 2010)." https://www.youtube.com/watch?v=5MGsDI0WJHQ (accessed August 2, 2017).

James, C. L. R. *The Black Jacobins: Toussaint L'ouverture and the San Domingo Revolution*. New York, NY: Vintage Books, (1963) 1989.

James, C. L. R. *A History of Pan-African Revolt*. Introduction by Robin D. G. Kelley. Oakland, CA: PM Press, Charlesh Kerr, Merlin Press, (1969) 2012.

James, C. L. R. *Toussaint Louverture: The Story of the Only Successful Slave Revolt in History, A Play in Three Acts*. Edited and with an Introduction by Christian Hogsbjerg. Foreword by Laurent Duboiss. Durham and London: Duke University Press, 2013.

Jessi, Furio. *Spartakus: The Symbology of Revolt*. Edited by Andrea Cavalletti. Translated by Alberto Toscano. London, New York, and Calcutta: Seagull Books, 2014.

Kaschl, Elke. *Dance and Authenticity in Israel and Palestine: Performing the Nation*. Leiden: Brill Academic Publishers, 2003.

Kelley, Robin D. G. *Freedom Dreams: The Black Radical Imagination*. Boston, MA: Beacon Press, 2002.

Kentridge, William. "MetCollects-Episode1/2014: William Kentridge on His Installation *The Refusal of Time*." http://www.metmuseum.org/metmedia/video/metcollects/the-refusal-of-time (accessed August 2, 2017).

Kentridge, William. *The Refusal of Time*. Paris: Editions Xavier Barral, 2012.

Kentridge, William. *Six Drawing Lessons*. The Charles Eliot Norton Lectures, 2010. Cambridge, MA and London: Harvard University Press, 2014. [The lectures are also online starting at http://mahindrahumanities.fas.harvard.edu/content/william-kentridge-drawing-lesson-one-praise-shadows (accessed October 12, 2016).]

Kershaw, Baz. *Theatre Ecology: Environments and Performance Events*. Cambridge: Cambridge University Press, 2009.

Latour, Bruno. "Trains of thought: The Fifth Dimension of Time and its Fabrication." http://www.bruno-latour.fr/sites/default/files/downloads/71-PIAGET-SWISS-VERSION.pdf (accessed October 12, 2016).

Lazzarato, Maurizio. "Grasping the Political in the Event: An Interview with Maurizio Lazzarato." *Inflexions:A Journal for Research-Creation*, no. 3 (October 2009). http://www.inflexions.org/n3_lazzaratohtml.html (accessed August 13, 2017). [PDF version at http://www.inflexions.org/n3_Grasping-the-Political-in-the-Event-Interview-with-Maurizio-Lazzarato.pdf (accessed August 13, 2017).]

Linebaugh, Peter, and Marcus Rediker. *The Many-Headed Hydra: Sailors, Slaves, Commoners, and the Hidden History of the Revolutionary Atlantic*. Boston, MA: Beacon Press, 2000.

Loveless, Natalie S. "The Materiality of Duration: Between Ice and Water Time." *Performance Research: A Journal of the Performing Arts* 19, no. 3 (2014): 129–36.

Lowe, Lisa. *The Intimacies of Four Continents*. Durham and London: Duke University Press, 2015.

Lubin, Alex. *Geographies of Liberation: The Making of an Afro-Arab Political Imaginary*. Chapel Hill, NC: University of North Carolina Press, 2014.

Meillassoux, Quentin. *After Finitude: An Essay on the Necessity of Contingency*. London and New York: Continuum, 2008.

Mirzoeff, Nicholas. *The Right to Look: A Counterhistory of Visuality*. Durham and London: Duke University Press. 2011.

MyEarbot. "Simultaneity – Albert Einstein and the Theory of Relativity." https://www.youtube.com/watch?v=wteiuxyqtoM (accessed August 1, 2017).

Negri, Antonio. *Time for Revolution*. London and New York: Continuum, 2003.

Nesbitt, Nick. *Caribbean Critique: Antillean Critical Theory from Toussaint to Glissant*. Liverpool: Liverpool University Press, 2013.

Nesbitt, Nick. "From Louverture to Lenin: Aimé Césaire and Anticolonial Marxism." *Small Axe* 19, no. 348 (November 2015): 129–44.

Nye, Coleman. "Cancer Previval and the Theatrical Fact." *TDR: The Drama Review* 56, no. 4 (2012): 104–20.

Nyong'o, Tavia. "The Scene of Occupation." *TDR: The Drama Review* 56, no. 4 (2012): 136–49.

Official Watts Prophets. "Dem Niggers Ain't Playin."™ https://www.youtube.com/watch?v=dEP6PIJubUI (accessed August 2, 2017).

Osborne, Peter. *The Politics of Time: Modernity and Avant-Garde*. London and New York: Verso Press, 1995.

Palladini, Giulia. "The Weimar Republic and its Return: Unemployment, Revolution, or Europe in a State of *Schuld*." In *Performances of Capitalism, Crises and Resistance: Inside/Outside Europe*. Edited by Marilena Zaroulia and Philip Hager. Baskingstoke: Palgrave Macmillan, 2015.

Perkinson, Robert R. *The Alcoholism and Drug Abuse Patient*. Thousand Oaks: Sage Publications, 2003.

Pezanoski-Browne, Alison. "Black to the Future: How Women in Pop are Carrying the Mantle of Afrofutruism." *Bitch Media*, February 16, 2014. https://www.bitchmedia.org/article/black-to-the-future-afrofuturism-feminism-music-janelle-monae-kelis-ebony-bones-theesatisfaction (accessed August 1, 2017).

Pfefferman, Naomi. “Star Choreographer Hofesh Shechter Discusses His “Political Mother.” *Jewish Journal*, October 18, 2011. http://jewishjournal.com/newspulse/97157/ (accessed August 1, 2017).

Phelan, Peggy. “On the Difference Between Time and History.” *Performance Research: A Journal of the Performing Arts* 19, no. 3 (2014): 114–19.

Ridout, Nicholas. *Passionate Amateurs: Theatre, Communism, and Love*. Ann Arbor, MI: The University of Michigan Press, 2013.

Rifkin, Mark. *Beyond Settler Time: Temporal Sovereignty and Indigenous Self-Determination*. Durham and London: Duke University Press, 2017.

Rockhill, Gabriel. “The Revolution Never Sleeps.” *Counterpunch*, April 15, 2016. https://www.counterpunch.org/2016/04/15/revolution-never-sleeps-nuit-debout-in-france-and-beyond/ (accessed July 19, 2017).

Ross, Christine. *The Past is the Present; it's the Future Too: The Temporal Turn in Contemporary Art*. New York and London: Bloomsbury Academic, 2012.

Ross, Kristin. *Communal Luxury: The Political Imaginary of the Paris Commune*. London and New York: Verso Press, 2015.

Schneider, Andrew. “You are Nowhere”. *Theater* 46, no. 2 (2016): 87–114.

Schneider, Rebecca. *Performing Remains: Art and War in Times of Theatrical Reenactment*. London and New York: Routledge, 2011.

Scott, David. *Omens of Adversity: Tragedy, Time, Memory, Justice*. Durham and London: Duke University Press, 2014.

Squad, Gob. “Before Your Very Eyes Interview with Gob Squad.” PublicTheaterNY, October 15, 2015. https://www.youtube.com/watch?v=WFAk3hikjgI (accessed August 7, 2017).

Stefano, Harney, and Fred Moten. “Fantasy in the Hold.” *The Undercommons: Fugitive Planning and Black Study*. Wivenhoe, New York, and Port Watson: Minor Compositions, 2013.

Stephens, Michelle Ann. *Black Empire: The Masculine Global Imaginary of Caribbean Intellectuals in the United States, 1914-1962*. Durham and London: Duke University Press, 2005.

Stephens, Paul, and Robert Hardwick Weston. “Free Time: Overwork as an Ontological Condition.” *Social Text* 26, no. 1 (94): 137–64.

Syms, Martine. “The Mundane Afrofuturist Manifesto.” http://martinesyms.com/the-mundane-afrofuturist-manifesto/ (accessed August 13, 2017).

Tadiar, Neferti X. M. “Life-Times of Disposability within Global Neoliberalism.” *Social Text* 31, no. 2 (Summer 2013): 19–48.

Tiedemann, Rolf. "Dialectics at a Standstill." In *The Arcades Project*. Translated by Howard Eiland and Kevin McLaughlin. Cambridge, MA and London: The Belknap Press of Harvard University Press, 1999.

TwoPlaysinRep. "A first look at WAITING FOR GODOT." https://www.youtube.com/watch?v=MUXtzkLTABI (accessed August 2, 2017).

Wagner, Matthew D. *Shakespeare, Theatre, and Time*. New York and London: Routledge, 2012.

Walsh, John Patrick. *Free and French in the Caribbean: Toussaint Louverture, Aimé Césaire and Narratives of Loyal Opposition*. Bloomington, IN: Indiana University Press, 2013.

West, Michael O., and William G. Martin. "Contours of the Black International: From Toussaint to Tupac." In *From Toussaint to Tupac: The Black International since the Age of Revolution*. Edited by Michael O. West, William G. Martin, and Fanon Che Wilkins, 1–44. Chapel Hill, NC: University of North Carolina Press, 2009.

Wickstrom, Maurya. "The Labor of Architecting." *TDR: The Drama Review* 54, no. 4 (Winter 2010): 118–35.

Wilder, Gary. *Freedom Time: Negritude, Decolonization, and the Future of the World*. Durham and London: Duke University Press, 2015.

Wilder, Gary. "Thinking with Aimé Césaire: Response by Gary Wilder." http://cesairelegacies.cdrs.columbia.edu/political-legacy/thinking-with-aime-cesaire/ (accessed August 1, 2017).

Yablonca, Eyal. "Israeli protesters dancing to Kesha—Tik Tok in Hebron (Rock the Casba)." https://www. youtube.com/watch?v=xVVte550dyU (accessed August 2, 2017).

Index

www.ingramcontent.com/pod-product-compliance
Ingram Content Group UK Ltd.
Pitfield, Milton Keynes, MK11 3LW, UK
UKHW020123290926
15471UKWH00003B/242